Organic Grower

THE AUTHOR

Dr. D. R. Singh is currently working as the Vice Chancellor, Chandrashekhar Azad Agricultural and Technological University, Kanpur. Born in 1964, his work experience is more than 28 years. During this time he has worked as Director of the Institute of Horticulture, Professor and Indian Council of Agricultural Research, New Delhi. Apart from this, various schemes have been embodied, guided, coordinated and implemented by him. He has been honored with 16 awards by various national and international organizations, including ICAR Fakhruddin Ali Ahmed, Lieutenant Governor's Commendation, Dr RS Paroda Award, Outstanding Scientist Award, Limca Book of National Records and Fellow awards etc. are prominent ones. Dr. Singh has developed 17 species of different horticultural crops as well as agricultural techniques of different crops and has successfully completed 14 projects funded by various institutions as principal investigator. He has published 206 research papers, 13 books, 50 book chapters and training bulletins at national and international level, as well as he is the chairman, expert and member of committees and scientific committees formed for various policy decisions. He is a renowned scientist in the field of horticulture at national and international level.

Organic Grower

– *Author* –

Prof. D.R. Singh

2022

Associated Publishing Company

A Division of

Astral International Pvt. Ltd.

New Delhi – 110 002

ISBN: 9789354615092

Published by : **Associated Publishing Company®**
A Division of
Astral International Pvt. Ltd.
– ISO 9001:2015 Certified Company –
4736/23, Ansari Road, Darya Ganj,
New Delhi-110 002
Ph. 011-43549197, 23278134
E-mail: info@astralint.com
Website: www.astralint.com

Preface

Organic farming is a technique in which cultivation of land takes place through sustainable manner using natural resources like air, water and soil. Organic farming does not necessarily mean that man-made chemicals are not used. In many organic farming methods, man-made fertilizers and pesticides are used not to improve crop production but to improve soil fertility. Which means that these chemicals are going directly into the soil and not into the food we consume? The difference between organic and inorganic farming is that in organic farming the chemicals help improve soil fertility and does not percolate into the crops but in inorganic farming, the chemicals degrade the soil fertility and make the soil barren after agricultural output and also results in these chemicals percolating into the food and ultimately into our food chains. Harmful chemicals such as zinc, lead and manganese that are present in pesticides and herbicides are consumed by human beings which can have grave dangers on our health.

Some of the major advantages of organic farming are that the production of healthy and quality agricultural output is relatively more as compared to inorganic farming. Organic farming is environment friendly since it helps to improve soil fertility and thereby the repeated usage of that particular patch of land for agricultural purposes is possible. This, in turn, helps in reducing deforestation caused due to agricultural practices. Other advantages of organic farming are that the pricing of organic output in the market is high which means that the profit margin for the farmer is also high. This sustained method of agriculture helps farmers in maintaining a constant and judicious balance of crop yield and saving our environment. Another biggest advantage of organic farming is it helps to prevent soil pollution.

This book is an effort to give simple and concise information about the aspects relevant to organic grower. The book consists of ten chapters. Efforts have been made to incorporate information on Undertake planning for organic farming, seed selection and treatment under organic farming, soil nutrient management under organic farming, weed control under organic farming, irrigation management under organic farming, integrated pest and disease management under organic farming, harvest and postharvest management under organic farming, undertake quality assurance and certification in organic farming, undertake business of organic farming, maintain health and safety at the workplace. The book provides you with comprehensive information on organic farming and related methods of farming. As valuable source of information to all those involved in organic farming including developmental personnel, policy makers, NGOs and progressive farmers.

I am grateful to all those persons as well as various books, manuals, periodicals, magazines, journals etc. that helped in the preparation of this book. In spite of the best efforts, it is possible that some errors may have occurred into the compilation and editing of the book. Further queries, constructive suggestions and criticisms for the improvement of the book are always welcome and shall be thankfully acknowledged.

Professor D. R. Singh

Contents

Chapter 1

Undertake Planning for Organic Farming

1.1 Seasonality of Various Crops

1.1.1 Organic Farming

- Organic farming system is being followed from ancient time.
- It is a method of farming system which mainly aims at cultivating the land and raising crops as to keep the soil alive and in good health using organic wastes like crop, animal and farm wastes, aquatic wastes and other biological materials along with beneficial microbes to release nutrients to crops for increased sustainable production in an eco-friendly pollution-free environment.
- For starting organic agriculture, crop grower needs to assess the type of produce he/she wants to grow by factoring in attributes like compatibility of the crop with soil, prevailing climatic conditions, popularity in the local market, *etc.*
- The methods of organic farming include biodynamic certification, home farming, weed and pest management, poly-cropping and crop diversity, microbe culturing, and vermicomposting and soil conservation with water harvesting.

Objectives

The objectives of organic farming are to:

- Produce food of high nutritional quality in sufficient quantities.

- Work in harmony with natural systems rather than seeking to dominate them.
- Encourage and enhance biological cycles in the farming system using microorganisms, soil flora, and fauna, plants, and animals.
- Maintain and increase the long-term fertility of soils.
- Use of renewable resources in locally organized agricultural systems
- Work in a closed system with regard to organic matter and nutrient elements
- Give all livestock, conditions of life that allow them to perform all aspects of their innate behavior.
- Avoid all forms of pollution that may result from agricultural techniques.
- Maintain the generic diversity of the agricultural system and its surroundings with the protection of plants and wildlife habitat.
- Allow agricultural producers an adequate return and satisfaction from work with a safe working environment.
- Consider the wider social and ecological impact of the farming systems.

Characters

The most important essential characters of organic farming include:

- Maximal but sustainable use of local resources
- Use of purchased inputs, only as complementary to local resources
- Increasing crop and animal diversity in the form of poly-cultures, agroforestry, integrated crop/livestock systems, *etc.* to minimize risk and improve ecological balance and economic stability
- Ensuring basic biological functions of the soil-water-nutrient-humus continuum
- Creating attractive overall landscape which gives satisfaction to local people

Advantages

- The crops are structurally and metabolically superior to other crops.
- These crops are less vulnerable to rotting or molding and can be stored for a protracted period.
- Input costs are low as compared to conventional farming
- Organic plants tolerate drought better and can grow in areas with meager irrigation facilities.
- Organic produces are tastier, as sugar content in them is richer, due to better quality nutrients by plants.
- Plant diseases, weeds, and pests are repelled or deterred by scientific crop rotation and other methods

- ☆ Yields from the same piece of land are higher.
- ☆ The customers are willing to pay more for organic products.
- ☆ It has secured income and rich returns on investment.

1.1.2 Season Requirement for Crops

- ☆ Crops are affected by soil type, rainfall, altitude, temperature, type and amount of nutrients required and amount of water needed.
- ☆ Different crops have different growing seasons in which local weather conditions such as rainfall and temperature allows normal crop growth.
- ☆ While each crop has a specific growing season which depends on its genetic adaptation, so growing seasons can be grouped in macro-environmental classes.
- ☆ It is seen that length of growing season of the crop varies from place to place and many crops requires the growing season of at least 90 days.
- ☆ The crop growing season in tropical regions where there is warm all the year-round, the growing season will last for the complete year.
- ☆ The crop growing season in temperate regions where there is warm summers and cold winters, the length of growing season of crops depends on temperature and such seasons will last for roughly eight months.
- ☆ The crop growing season is also affected by elevation or height above sea level, as higher elevations mostly have cold temperatures.
- ☆ The crop growing season in temperate regions, is calculated by average number of days, between last frost in spring and first severe frost in autumn.
- ☆ The growing season for crops can also be determined by average number of days that the temperature rises high enough for a particular crop to sprout and grow, which varies as per the crop.

1.2 Appropriate Times for Planting by taking Soil

1.2.1 Soil

- ☆ It is a critical resource that manages the crop and led to improvement or degradation of its quality.
- ☆ It is a living system and its fertility is main to agricultural productivity.
- ☆ The maintenance of soil fertility is the primary step in any agricultural system.
- ☆ Soils deficient in the nutrient cannot support either crop production or active populations of beneficial microorganisms, which are essential for a productive soil.

- Soil conditions for planting different organic crops include soil texture, color, organic matter content, pH, fertility levels which can be improved or altered by the addition of organic fertilizers, such as compost or manure.
- To soil organic carbon content in wind-eroded sandy soils is reported 50 per cent less than that in non-eroded sandy loam soils.
- Temperature and rainfall have a significant influence on organic carbon levels in the soil which takes place due to their controlling effect on plant growth and microbial activities.
- Better plant growth gives higher input of carbon to the soil, while higher microbial activities cause greater depletion of soil carbon due to more decomposition of soil organic carbon including the residues.
- The difference between the magnitude of their individual effect on addition and depletion of carbon determines the net balance of carbon in the soil which is universally evident.
- The low soil carbon status is attributed to high decomposition rates that decompose all organic carbon, but high rainfall and low-temperature conditions are conducive to the accumulation of organic carbon in soils, while high temperatures and low rainfall conditions are not conducive for the build-up of carbon stock in soil.

1.2.2 Planting Time of Crops

- Organic farming takes time and requires more attention as compared to conventional farming.
- There are certain techniques that need to be considered as per the crop and farm conditions.
- Different crops have different planting time and seasons.

For potatoes planting on hills and in plains, the season and planting time will be:

Hills

- Summer: From March to April
- Autumn: From August to September
- Irrigation: From January to February

Plains

- Planting time is from October to November

For pulse crops, the planting time will be:

- Early Kharif: from August to October
- Summer: March to June

- Kharif: From June to September or August to October
- Rabi: From October to January or from August to October

1.3 Climatic Conditions

1.3.1 Introduction

- Climate is one of the main determinants of agricultural production and change in climate causes variability in agricultural production.
- Depending on climatic conditions, different crops will be harvested at different points throughout the year.
- As climate pattern shifts, changes in distribution of plant diseases and pests will have adverse effects on agriculture.
- If a crop is grown in a climate to which it is not suited, it is likely to produce low yields and be more susceptible to the pest and diseases which creates the need to use agrochemicals to fertilize the crop and control pest and diseases.
- Organic farming serves as low-risk farming strategy with reduced input costs, lower risks with partial or total crop failure as a result of extreme weather events or changed conditions in the wake of climate change and variability.
- Organic farms often sustain higher species diversity and cultivate locally adapted varieties which improve resilience of agro ecosystems against adverse climate conditions, such as extreme weather events.
- The successful organic farmer learns to grow the crops and varieties which are suited to the local conditions.
- Organic farms often sustain higher species diversity and cultivate locally adapted varieties which improve resilience of agro ecosystems against adverse climate conditions, such as extreme weather events.
- The wind is an important climatic parameter which causes soil erosion under arid condition.

Organic Farming in Low Rainfall Areas

- In low rainfall areas, multi-component farming systems are prevalent that cover annuals, perennials, and animal components.
- Such systems have minimal or no external inputs and are based on the recycling of local resources and help in spreading the risk of rainfall uncertainty.
- These systems have sustained production reasonably under prevailing climatic uncertainties, their productivities are very low which can be attributed to inefficient use of local resources.

- Under such conditions use of synthetic inputs is risk-prone and uneconomic.
- In the last 4-5 decades, many attempts have been made to improve the productivity of traditional systems using synthetic external inputs like fertilizers, pesticides, weedicides, but success was limited to good rainfall years only.
- In low rainfall areas due to high risk linked with crop production, the use of agrochemicals is very less and farmers are practicing organic farming for ages.
- Such a system, despite its low productivity as compared to the conventional system, has established itself as highly efficient in terms of resource recycling and providing better food and economic sustainability in an arid region.
- Many promising crops for production under the organic systems are cumin (spice), psyllium (medicine), sesame (oilseed), cluster bean (gum), moth bean (papad, bhujia), *etc.*

Required Climatic Conditions for Food Grains

Food Grains	*Climatic Condition*
Rice	**Temperature:** 22 -32 degree Celsius **Rainfall:** 150-300 cm **Soil Type:** Deep clayey and loamy soil
Wheat	**Temperature:** 10-15 degree Celsius (Sowing time) **Temperature:** 21-26 degree Celsius (Ripening and Harvesting **Rainfall:** 75-100 cm **Soil Type:** Well-drained fertile loamy and clayey loamy
Millets	**Temperature:** 27-32 degree Celsius **Rainfall:** 50-100 cm **Soil Type:** They are less sensitive to soil deficiencies. They can be grown in inferior alluvial or loamy soil
Grams	**Temperature:** 20-25 degree Celsius (Mild cool and Dry Climate **Rainfall:** 40-45 cm **Soil Type:** Loamy Soil
Sugar Cane	**Temperature:** 21-27 degree Celsius **Rainfall:** 75-150 cm **Soil Type:** Deep rich loamy soil

Food Grains	*Climatic Condition*
Cotton	**Temperature:** 21-30 degree Celsius **Rainfall:** 50-100 cm **Soil Type:** Black soil of Deccan and Malwa Plateau. However, it also grows well in alluvial soils of the Sutluj-Ganga plain and red and laterite soils of the peninsular region
Oilseeds	**Temperature:** 20-30 degree Celsius **Rainfall:** 50-75 cm **Soil Type:** Well drained light sandy loams, red, yellow and black soils are well suited for its cultivation.
Tea	**Temperature:** 20-30 degree Celsius **Rainfall:** 150-300 cm **Soil Type:** Well drained, deep friable loamy soil.
Coffee	**Temperature:** 15-28 degree Celsius **Rainfall:** 150-250 cm **Soil Type:** Well drained, deep friable loamy soil.

1.4 Advantages of Crop Rotation

1.4.1 Crop Rotations

- ☆ Growing the same crops in the same field year after year reduces soil fertility and encourages build up of pests, diseases and weeds in soil.
- ☆ Crops should be moved to a different area of land each year, and should not returned to original site for several years.

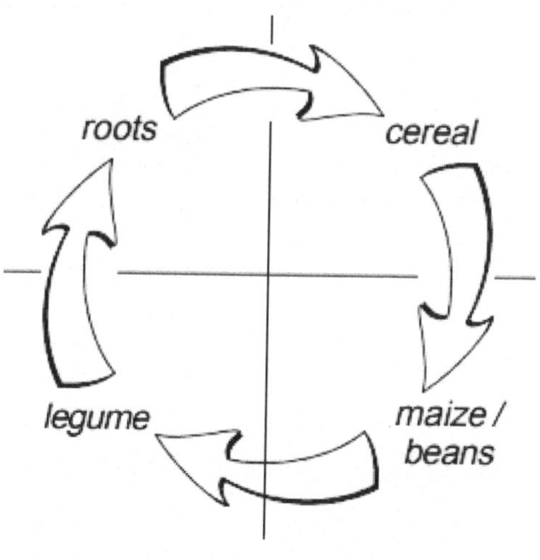

- ☆ For vegetables, minimum 3-4 years of rotation is usually recommended.
- ☆ Crop rotation means having times where soil fertility is being built up and times where crops are grown which remove nutrients.
- ☆ Crop rotation helps many natural predators to survive on farm by providing diverse habitats and sources of food.

1.4.2 Features

- ☆ Organic farming relies on good, natural soil health, so crop rotation is big part of process.
- ☆ Rotating crops in the field will keep the soil balanced in the field so that certain nutrients don't get depleted, as various crops need different nutrients that leave other crops behind.
- ☆ Crop rotation is important to reduce pressure from insects, weeds, while supplementing nutrients in soil.
- ☆ Trap crops that are planted between main crops also keep the soil rich and helps in avoiding erosion.

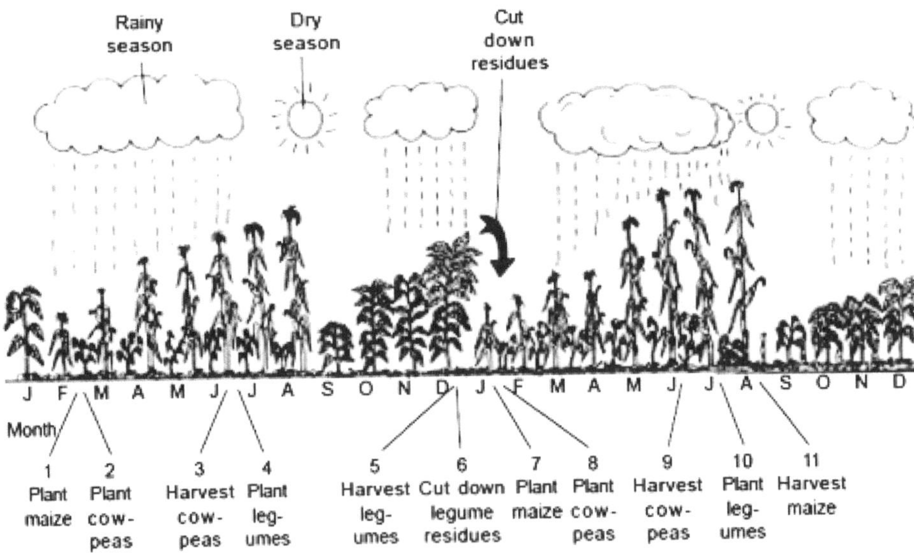

2-Year Crop Rotation of CEREALS, COWPEAS AND LEGUMES.

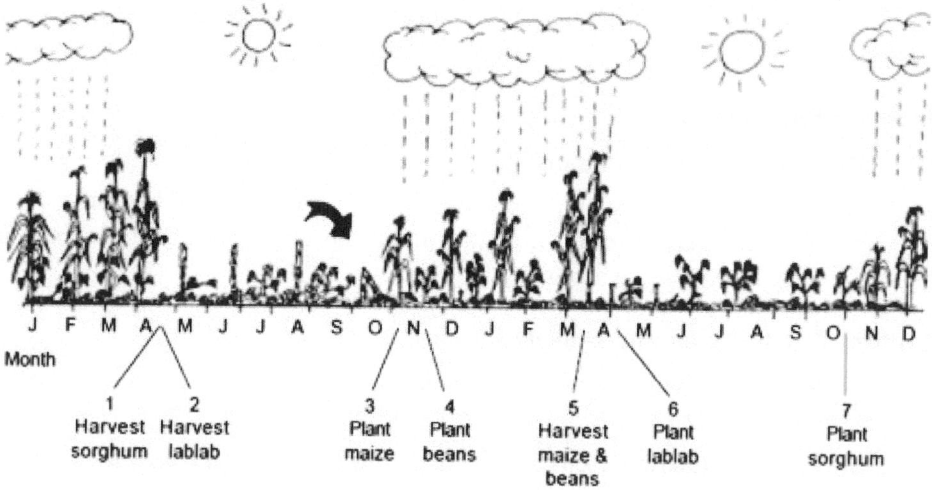

2-Year Crop Rotation of MAIZE, BEANS, SORGHUM AND LABLAB.

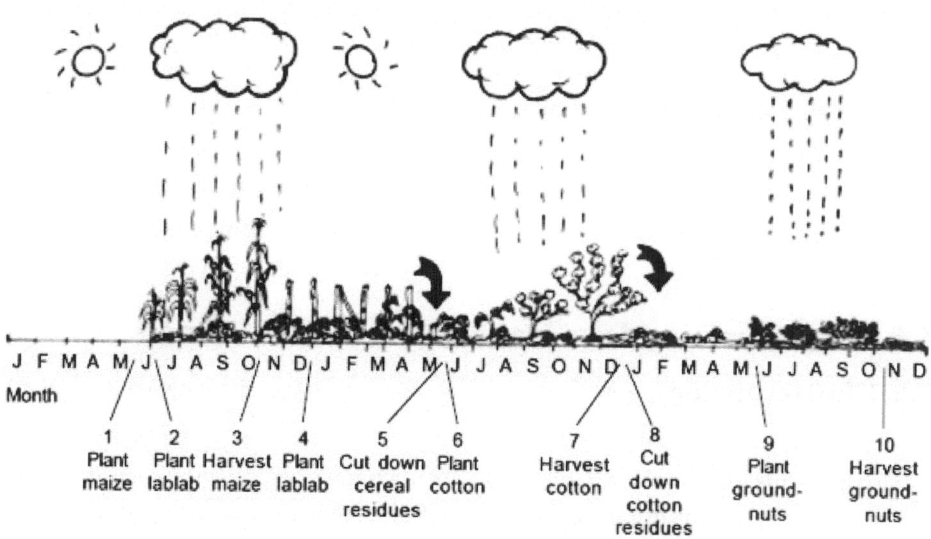

3-Year Crop Rotation of CEREALS, COTTON AND GROUNDNUT.

1.4.3 Advantages

The advantages of crop rotation:

Improve Soil Structure

- ☆ Many crops have strong, deep roots, which can break up hardpans, and tap moisture and nutrients from deep in soil while others have many fine, shallow roots which tap nutrients near surface that binds the soil to form many tiny holes for air and water to get inside the soil.

Increase Soil Fertility

- Legumes like groundnuts/beans fix nitrogen in soil.
- In legume green parts and roots rot, nitrogen can be used by other crops like maize which results in higher and more stable yields without the use to apply expensive inorganic fertilizer.

Control Weeds, Pests and Diseases

- Planting similar crop season after season encourages certain weeds, insects and diseases.
- Planting different crops breaks their life cycle and prevents them from multiplying.

Produce different Output

- Growing a mix of grain, beans, vegetables and fodder means more varied diet and more types of produce to sell.

Used for Ploughing the Soil

- It helps aerate the soil, recycles nutrients and helps in controlling weeds, pests and diseases.
- Intercropping, strip cropping and relay cropping bring many advantages as rotation.

1.5 Advantages and Disadvantages of Intercropping

1.5.1 Intercropping

- It is a practice of growing two or more crops in close proximity.
- In this, two or more cash crops are grown together, growing cash crop with cover crop or other non-cash crop that provide benefits to primary crop.
- It requires extra management to keep competition between intercropped species in balance.
- The idea of intercropping is to get highest potential yield off a particular field by maximising the potential of resources present at given time.
- It is a system that has been used traditionally by small-scale farmers in tropical climates where cereal and legume intercropping combination is common.
- It has four general spatial arrangements as mixed intercropping with no distinct row arrangement, strip intercropping where crops are grown

in strips that are wide enough to separate or narrow enough to allow interaction between them, relay intercropping where two or more crops are grown during differing parts of life cycles and row intercropping where at least one crop is planted in rows.

- ☆ In this when two or more crops are growing together, each must have adequate space to maximize cooperation and minimize competition between them which needs spatial arrangement, plant density, maturity dates of crops being grown and plant architecture.

1.5.2 Approach

There are different approaches to intercropping such as:

Mixed Intercropping

- ☆ Two or more crops are planted in a mix without a distinct row arrangement.
- ☆ This type of intercropping is common in intensive subsistence systems in subtropical and tropical areas across the world.
- ☆ It is often used where fields are relatively small and field operations involve mostly manual labor.
- ☆ It ensures that ground is covered, and therefore weeds are suppressed.
- ☆ The system can be very complex with number of crops produced together.

Row Intercropping

- ☆ In this, two or more crops are planted in distinct rows.
- ☆ The productivity in such intercropping is influenced by specific combination of crops, their spacing and planting dates.
- ☆ The intercropping is ideal for shade loving plants like groundnuts and cowpeas.

Relay Intercropping

- ☆ In this, two or more crops are grown at same time as part of life cycle of each, where other crop is sown after first crop gets well established, but before the first crop reaches its harvesting stage.
- ☆ Such type of intercropping is practiced in regions where growing season is too short to allow two sequential crops.

Strip Intercropping

- ☆ In this, two or more crops are grown at same time in separate strips which is wide enough apart for independent cultivation.
- ☆ It includes contour strip cropping and field strip cropping, where contour strip cropping follows a layout of definite rotational sequence and tillage

is held closely to exact contour of field while field strip cropping has strips with uniform width that follows across general slope of land.

- It is ideal for vegetables.
- It is easier to control pest outbreaks in one strip as such method ensures that crops of same family are not too near to each other.

1.5.3 Advantages of Intercropping

More Income and More Yield

- Intercropping offers greater financial returns for the farmer.
- Even if crop growers are growing some produce for their own family or just as part of hobby, they will have multiple types of produce, which is always good outcome.
- It helps crop growers to use the same land and yield more as well as diverse produce to generate more income without really taking up any major expenditure.
- The infrastructure available or land used in this remains the same.

Insurance Against Crop Damage

- Intercropping provides an insurance which the crop growers require, especially when region is vulnerable to weather extremes.
- Drought, torrential rain, hurricanes or cyclones and various other weather elements can affect the yield of given year or season.
- Having diverse yields allows crop growers to have some income even if primary crop gets damaged or doesn't yield as required.

Optimum Use of Soil

- Intercropping makes the most of available soil.
- If anything is grown on farmland, crop tends to absorb as much water and nutrients as it needs.
- There could be more nutrients in soil under the crops and around, so soil and nutrients can be used, by different varieties of crops.
- Intercropping also averts soil runoff and can prevent the growth of weeds.

Good for Primary Crops

- Intercropping is good for primary crops.
- The secondary crops can provide shelter and even protect the primary crops.
- Intercropping allows crop growers to grow cash crops or any crop that will actually supplement the primary crop in some way.

1.5.4 Disadvantages of Intercropping

Poor Yields

- Intercropping can lead to poorer yields.
- In this, crops may not be compatible.
- The crops may actually compete for the same nourishment and of course water, which may lead to an unmanageable conflict.
- In this, it is possible that both crops don't yield enough produce.

Costly and Complicated

- Intercropping costs more money upfront.
- There is a need for more fertilizers and water.
- Harvesting is also more complicated and when something goes wrong with either crop, then other crop will also get damaged.

1.6 Types of Plant to be Intercropped

1.6.1 Introduction

- Intercropping is a practice of growing two or more crops in the same row or bed or in rows or strips that are close enough for biological interaction.
- Almost any type of crop or combination of crops can be used for intercropping.
- Organic grower should select crops that are locally grown and well adapted to the climatic conditions.
- Seeds or other plant materials can be obtained from other farmers, or local markets or from specialized seed producers.
- There are many different types of species that can be used for intercropping like annuals, *e.g.* cereals and legumes, perennials including trees, or a mixture of the two.

1.6.2 Types of Intercropped Crops

The crops which are usually intercropped by organic crop grower are:

Crop	*Intercrop With*
Asparagus	Tomato, Parsley, Basil
Bush Beans	Potato, Cucumber, Corn, Strawberry, Celery, Summer Savory
Pole Beans	Corn, Summer Savory, Radish
Cabbage Family	Aromatic Herbs, Celery, Beets, Onion Family, Chamomile, Spinach, Chard

Crop	*Intercrop With*
Carrots	Radishes, Lettuce, Rosemary, Onion Family, Sage, Tomato
Celery	Onion, Cabbage Families, Tomato, Bush Beans, Nasturtium
Corn	Potato, Beans, Pumpkins, Cucumber, Squash
Eggplant	Beans, Marigold
Lettuce	Carrots, Radish, Strawberry, Cucumber
Onion Family	Beets, Carrots, Lettuce, Cabbage Family, Summer Savory
Parsley	Tomato, Asparagus
Potato	Beans, Corn, Cabbage Family, Marigolds, Horseradish
Pumpkins	Beans, Corn, Marigold
Radish	Carrots, Nasturtium, Lettuce, Cucumber
Spinach	Strawberry, Beans
Squash	Nasturtium, Corn, Marigold
Tomato	Onion Family, Nasturtium, Marigold, Asparagus, Carrots, Parsley, Cucumber
Turnip	Aromatic Herbs, Celery, Beets, Onion Family, Chamomile, Spinach, Chard

1.7 Effects of Specific Chemicals on Health and Environment

1.7.1 Exposure of Chemical/Fertilizers

- ✰ Exposure to chemicals/pesticides can occur in many ways as farmers/growers/workers and people residing in surrounding areas can be exposed to pesticides by treatment of crops, plants and grain stores.
- ✰ Rural residents living next door to fields can be exposed to pesticide drift.
- ✰ Exposure can also occur in forestry, professional and domestic pest control, by treatment of wood with preservatives, treatment of boat hulls with anti-fouling agents and treatment of livestock with anti-parasitic preparations.

1.7.2 Risk

There are various risk associated with use of chemical/pesticides such as:

1. **Risk to Humans**
   - ✰ Pesticides have linked with deleterious effects on human health and on environment.
   - ✰ Children exposed to pesticides have increased rates of leukemia and brain cancer.

- ☆ Pregnant women with exposure to pesticides lead to higher miscarriage rates.
- ☆ Pesticides damage the lungs and nervous system of humans.
- ☆ Pesticides are acute toxic as they can lead to harmful or lethal effects after single episode of ingestion, inhalation or skin contact.
- ☆ A substance is considered carcinogenic when there is evidence that it can cause cancer.

2. Risk of Chronic Toxicity

- ☆ Pesticides can cause harmful effects over extended period, following repeated or continuous exposure at low levels.
- ☆ Low doses don't always cause immediate effects, but can lead to serious illnesses.
- ☆ Long term pesticide exposure has been linked to development of Parkinson's disease; asthma; depression and anxiety; attention deficit and hyperactivity disorder and cancer, including leukemia and non-Hodgkin's lymphoma.

3. Risk to Wildlife

- ☆ The impacts of pesticides on wildlife are extensive and expose animals in urban, suburban, and rural areas to unnecessary risks.
- ☆ Wildlife can be impacted by pesticides by their direct or indirect application like pesticide drift, secondary poisoning, and runoff in local water bodies or groundwater contamination.
- ☆ In birds, exposure to certain pesticides can obstruct singing ability, which makes difficult to attract mates and reproduce.
- ☆ In nature, pesticides pollute the air, water and ground with which plant and animal life die or become sick and malformed.

1.8 Characteristics of Good Agricultural Practices (G.A.P)

1.8.1 Good Agricultural Practices

- ☆ Good Agricultural Practices refers to farming methods that reduce the likelihood of contaminating produce.
- ☆ Implementation of practices that address water quality, manure and compost use, worker health and hygiene and contamination from wildlife, domestic animals and livestock.
- ☆ It has evolved in recent years in the context of rapidly changing and globalizing food economy and as a result of concerns and commitments of a wide range of stakeholders regarding food production and security, food safety and quality, and environmental sustainability of agriculture.

- The stakeholders represent actors from supply dimension (farmers, farmers' organizations, workers), demand dimension (retailers, processors, and consumers) and those institutions and services (education, research, extension, input supply) that support and connect demand and supply and who seek to meet specific objectives of food security, food quality, production efficiency, livelihoods and environmental conservation in medium and long term.
- There are four pillars of Good Agricultural Practices that cover economic viability, environmental sustainability, social acceptability, and food safety and quality in private and public-sector standards.
- The concept of Good Agricultural Practices serves as a reference tool for deciding, at each step in the production process, on practices and/or outcomes that are environmentally sustainable and socially acceptable.

Objectives

The objective of Good Agricultural Practices codes, standards and regulations covers varying degree that:

- ensure safety and quality of produce in the food chain.
- capture new market advantages by modifying supply chain governance.
- There are four pillars of Good Agricultural Practices that cover economic viability, environmental sustainability, social acceptability, and food safety and quality in private and public-sector standards.
- create new market opportunities for farmers and exporters in developing countries

1.8.2 Good Agricultural Practices Related to Crop Protection

- Using resistant cultivars and varieties.
- Crop sequences, associations and cultural practices.
- Biological prevention of pests and diseases.
- Maintain regular and quantitative assessment of balance status among pests and diseases and beneficial organisms of all crops.
- Adopt organic control practices where and when applicable.
- Applying pest and disease forecasting techniques where available.
- Determining interventions by following consideration of methods with short and long-term effects on-farm productivity and environmental implications to allow minimizing agrochemicals and promote Integrated Pest Management.
- Storing and using agrochemicals as per legal need of registration for each crop, rates, timings, and pre-harvest intervals.

- ☆ Ensuring that agrochemicals are applied by specially trained and knowledgeable persons.
- ☆ Ensuring that equipment used for handling and application of agrochemicals complies with established safety and maintenance standards.
- ☆ Maintaining accurate records of agrochemical use.
- ☆ Identifying the GAP in each protection method.

1.9 Natural, Organic and Zero Till Farming

1.9.1 Natural Farming

- ☆ It is a system where laws of nature are applied to agricultural practices.
- ☆ It works along with the natural biodiversity of the farmed areas, encourages the complexity of living organisms both plants and animals that shape each particular ecosystem to thrive along with food plants.
- ☆ It is a chemical-free and poison-free farming method.
- ☆ It discourages farmers from using any chemical fertilizers, pesticides on plants and in all agricultural practices.
- ☆ It encourages farmers to use local breeds of seeds, and native varieties of vegetables, grains, pulses, and other crops.
- ☆ In this, neither chemical nor organic fertilizers are added to the soil and no external fertilizers are added to soil or given to plants.
- ☆ In this, decomposition of organic matter by microbes and earthworms encourages on the soil surface itself, which gradually adds nutrition in the soil, over the period.
- ☆ In this, there is no plowing, no tilting of soil and no fertilizers and no weeding is done in natural ecosystems.

1.9.2 Organic Farming

- ☆ It involves a chemical-free and less poison-free farming methods.
- ☆ It discourages farmers from using any chemical fertilizers, pesticides on plants and in all agricultural practices.
- ☆ It encourages farmers to use local breeds of seeds, and native varieties of vegetables, grains, pulses, and other crops.
- ☆ It promotes nonchemical and homemade pest control methods.
- ☆ In organic farming, organic fertilizers and manures like compost, vermicompost, cow dung manure, *etc.* are used and added to farmlands from external sources.
- ☆ It still needs basic agro practices like plowing, tilting, mixing of manures, weeding, *etc.* to be performed.

- It is expensive due to the requirement of bulk manures and has an ecological impact on surrounding environments.
- Protect soil quality using organic material and encourage biological activity.
- In this, weed and pest control based on methods like crop rotation, biological diversity, natural predators, organic manures and suitable chemical, thermal and biological intervention.
- In this, their results care for a large environment and conservation of natural habitats and wildlife.

1.9.3 Zero Tillage Farming

- It involves less soil erosion from wind and water as mulch cover of previous crops covers the soil.
- Protect soil quality using organic material and encourage biological activity.
- It is more fertile and resilient soils.
- It has less moisture evaporation.
- It has low fuel and labor costs as there are fewer passes across the field.
- In this, the upfront costs of zero tillage equipment are high.
- It has less soil compaction.

1.10 Benefits and Trade-Offs in Organic Farming

- Organic agriculture is a mechanism which reduces greenhouse gases as no chemical nitrogen fertilizers are used and stores carbon in soil and plant biomass by building organic matter.
- It encourages agro-forestry and inhibits the clearance of primary ecosystems and minimizes energy consumption by eliminating the energy required to manufacture synthetic fertilizers and uses internal farm inputs.
- The mitigation of climate change in organic farming includes improved crop and grazing land management with the restoration of soil and degraded land.
- Soil carbon sequestration is responsible for many mitigation potentials which play a role in reducing erosion and minimizing impact on native ecosystems.
- It has less moisture evaporation.
- It has low fuel and labor costs as there are fewer passes across the field.
- Organic agriculture promotes human health as organic food plays an important role due to its high nutritional quality and quantity and reduction of harmful chemicals.

- Organic agriculture is linked with lower yields, as it is profitable as conventional agriculture, if not more so.
- The financial benefits of organic agriculture covers lower cash expenses, high net revenue, and the low breakeven point.
- Organic agriculture compares favorably with conventional agriculture when comparison includes full cost accounting of environmental and health harms or benefits.

1.11 Common Methods of Preparation of Organic Farming Inputs

1.11.1 Organic Farm Input

- In agriculture, input is any substance used by the producer for pest control or for soil fertility management.
- Organic inputs such as various types of composts, biofertilizers, *etc.* are not only important in organic farming but are also of prime importance in integrated agriculture with balanced use of fertilizers for the sustenance of soil fertility.
- Organic inputs include composts, vermicompost, coir pith composting, liquid manure and biofertilizer.
- Input can be natural or synthetic.
- Natural inputs include substances that are created through the biological process which naturally occurs like vinegar.
- Many natural inputs that are prohibited are products derived by genetic engineering and contaminated organic materials like conventional cottonseed meal having pesticides and leather meal containing heavy metal chromium.
- Inputs for farms help in build humus, supply nutrients and build microbial activity and diversity to help with disease and crop health.
- The main role of organic inputs is to be used either as fertilizer for crops or as a soil amendment.
- Organic inputs enable a stable level of organic matter in soil and improve soil structure, stimulate biological activity, increase water retention, and facilitate plant health.

1.11.2 Organic Farm Input Methods

Methods of Preparation of Compost/Vermicompost

Vermicasting

- It is a natural, odorless, aerobic process, much different from traditional composting.

- Vermicast nutrient content varies with earthworm feed type, but feeding waste to earthworms does cause nitrogen mineralization, followed by phosphorus and sulfur mineralization after ejection.
- It can be done on a small scale by homeowners with household organic wastes and on large-scale by farmers with manure or by food industry using organic wastes like fruit and vegetable cull materials.
- For preparing compost, the grower should use stacked bin system which should have appropriate drainage and aeration, container walls, bottom, and sidewalls.
- The vermicast bins have a few holes at the bottom for drainage with a screen-type floor surface that allows drainage, unlike common bedding materials.
- It has a screen size of 4 mm (5/32 in.) which allows to keep most waste particles in it.
- Turning organic wastes into casts takes 22–32 days as per the density of waste and earthworm maturity.
- It does not need curing, but fresh casts undergo 2 weeks of nitrification where ammonium transforms to nitrate for plants to uptake.
- In this, organic materials which meet earthworm's feed preferences include material density of 350–650 g/L.
- While doing, the worms should not be crowded, so ideal stocking density is 150 earthworms/L of wastes.
- In this, earthworms ingest 75 per cent of body weight/day as 0.2 g worm eats about 0.15 g/day and if it tries to escape, then it shows that something is wrong with their feed or environment as they are not allowed to leave about 1 week to migrate from finished vermicast to fresh waste.

Methods of Preparation of Biofertilizers

Biofertilizer

- They are carrier-based preparations with effective strains of microorganisms like bacteria, fungi, and algae alone or in combination.
- They contain microorganisms that promote the adequate supply of nutrients to host plants and ensure proper development of growth and regulation in their physiology.
- They improve soil texture, pH, WHC and other properties of soil and produce plant growth by promoting substances IAA, amino acids, vitamins, *etc.*

- They contain 3.5 per cent-4 per cent nitrogen, 2 per cent-2.5 per cent phosphorus and 1.5 per cent potassium.
- On being applied on seed, seedling roots or in soil fix atmospheric nitrogen or solubilize insoluble phosphate in soil and make them available to crop plants.
- Biofertilizers are bio inoculants/culture, microbial inoculants or teeka.
- For the production of good and efficient bio-fertilizer, an efficient nitrogen-fixing strain is required with inoculum.

Production of Biofertilizer

- The production of Bio-fertilizer involves culturing of microorganisms, processing of carrier material and mixing carrier and broth culture and packing.
- Biofertilizers are carrier-based preparations with an efficient strain of nitrogen-fixing or phosphate solubilizing microorganisms prepared in suitable medium and mass-produced in fermentor with proper aeration, temperature, growth conditions and without any undesired microbial contaminations.
- They are formulated as carrier-based inoculants where organic carrier materials are effective for the preparation of bacterial inoculants.
- Peat soil, lignite, vermiculite, charcoal, press mud, farmyard manure and soil mixture can be used as carrier materials.
- Here, neutralized peat soil/lignite are better carrier materials for bio-fertilizer production which is cheaper, locally available, inert, having high WHC and organic content.
- The bacterial culture drawn from fermentor is added to the sterilized carrier and mixed well by the manual or mechanical mixer and sealed at room temperature.

1.12 Basic Financial Plan for Organic Farming Breakeven

1.12.1 Financial Plan

The financial plan of organic farming covers following:

- Required Cost of Start-Up
- Profit and Loss
- Cash Flow
- Balance Sheet
- Financial Ratios

Start-Up Costs

The following chart summarizes start-up expenses:

Start-Up Expenses	Start-Up Assets
Operating Capital	Real Estate
Salaries and Wages	Buildings
Insurance Premiums	Leasehold Improvements
Beginning Inventory	Equipment
Legal and Accounting Fees	Furniture and Fixtures
Rent Deposits	Vehicles
Utility Deposits	Other Fixed Assets
Supplies	
Advertising and Promotions	
Licences	
Other Initial Costs	
Working Capital (cash on hand)	
Total Start-Up Expenses	

Source and Use of Funds

To date, owners need to invest certain percent of total costs involved which can be arranged from:

Source and Use of Funds
Sources of Funds
Owner's Injection
Commerical Loan
Commerical Mortgage
Inventory and Working Capital Loan
Total Source of Funds
Use of Funds
Fixed Assets
Operating Capital
Total Use of Funds

Break-Even Analysis

Grower needs to estimate total fixed costs which will be involved in the project.

The variable costs such as salaries, fuel charges, postage, repairs, maintenance and supplies needs to be calculated per unit.

Units are considered full shares for analysis purposes and do not consider half shares.

- ☆ Monthly Revenue Break-even
- ☆ Assumptions
- ☆ Average Percent Variable Cost
- ☆ Estimated Monthly Fixed Cost

Based on assumption of average share price, breakeven revenue will be obtained as shown:

NET UNITS	NET REVENUE	FIXED COST	VARIABLE COST	TOTAL COST	TOTAL PROFIT

Graph that follows:

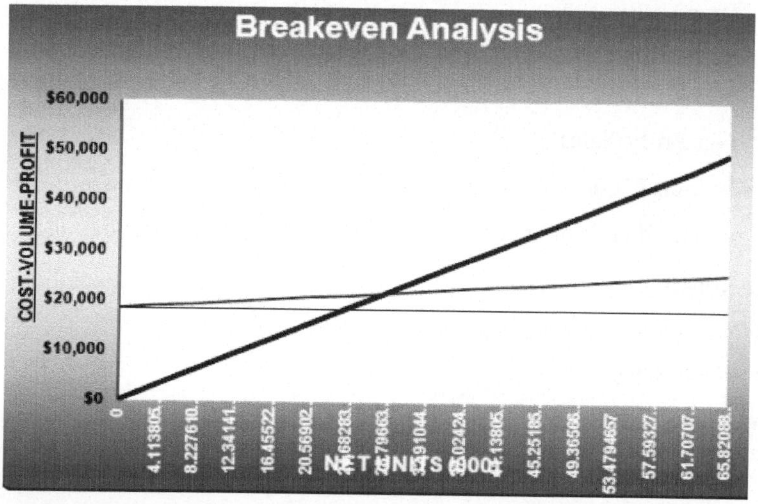

Projected Profit and Loss

The projected profit and loss can be obtained by using following details:

Proforma Profit and Loss

- Sales
- Direct Cost of Sales
- Other Production Expenses

Total Cost of Sales

- Gross Margin
- Gross Margin per cent
- Expenses
- Payroll
- Sales and Marketing and Other Expenses
- Depreciation
- Leased Equipment
- Utilities
- Insurance
- Rent
- Payroll Taxes
- Other
- Total Operating Expenses
- Profit Before Interest and Taxes
- EBITDA
- Interest Expense
- Taxes Incurred
- Net Profit
- Net Profit/Sales

Projected Cash Flow

The projected cash flow can be obtained by using following details:

Cash Received

- Cash from Operations
- Cash Sales
- Cash from Receivables

Subtotal Cash from Operations

- ☆ Additional Cash Received
- ☆ Sales Tax, VAT, HST/GST Received
- ☆ New Current Borrowing
- ☆ New Other Liabilities (interest-free)
- ☆ New Long-term Liabilities
- ☆ Sales of Other Current Assets
- ☆ Sales of Long-term Assets
- ☆ New Investment Received

Subtotal Cash Received

- ☆ Expenditures
- ☆ Expenditures from Operations
- ☆ Cash Spending
- ☆ Bill Payments

Subtotal Spent on Operations

- ☆ Additional Cash Spent
- ☆ Sales Tax, VAT, HST/GST Paid Out
- ☆ Principal Repayment of Current Borrowing
- ☆ Other Liabilities Principal Repayment
- ☆ Long-term Liabilities Principal Repayment
- ☆ Purchase Other Current Assets
- ☆ Purchase Long-term Assets
- ☆ Dividends

Subtotal Cash Spent

- ☆ Net Cash Flow
- ☆ Cash Balance

1.13 Common Transition Practices

1.13.1 Organic Transition

- ☆ Transition from conventional to organic farming needs many changes which involve changes in the mindset of crop growers.
- ☆ Transitioning farmers spend too much time worrying about replacing the synthetic input with allowable organic product rather than just considering

management practices as per the preventative strategies which involve new entrants like:

Understanding the Basics of Organic Agriculture and Farming Standards

- ☆ As organic production systems are knowledge-based, so new entrants and transitional producers should become familiar with the sound and sustainable agricultural practices.
- ☆ Transitional producers should be prepared to read appropriate information, conduct their own trials and participate in formal and informal training events.
- ☆ Switching from conventional to organic farming is more than substituting synthetic materials to organic allowed materials.
- ☆ Organic farming is a holistic system that relies on sound practices as focused on preventative strategies.
- ☆ As there are often few organic remedies for organic producers for certain problems, prevention is considered as the main element in organic production.

Identifying Useful Resources

- ☆ Existing crop growers are helpful in sharing valuable technical information as a good mentor should able to show transitional producers with knowledge, practical experience and should suggest appropriate reading materials.
- ☆ Mentors should identify some of the important challenges that transitional farmers will confront and should help source production materials that are otherwise difficult to find.
- ☆ Producers should also contact agrologists, veterinarians and other agricultural and financial consultants, to learn ways in order to improve current farming practices.
- ☆ As the internet is valuable source of information for new organic growers where a broad range of reading materials are present for growers to access valuable resources and establish good working contacts.

Planning of Transition

- ☆ Organic crop grower should develop a transitional plan with clear and realistic goals which should clearly identify various steps involved in making the transition to organic that should cover realistic timeframes.

Identifying Strengths and Weaknesses

- ☆ There are certain ways through which crop growers can address their weaknesses while building on strengths.

- The business side of the transitional plan should have many year budgets and effective/realistic marketing strategies.
- It is required for crop grower to list their expenses comprehensively which should cover prerequisites to begin the transition like mechanical weeding equipment, specialized composting equipment and applicators, additional handling equipment dedicated to organic products, and processing equipment.
- As the demand for organic products is growing, growers need to make sure, that they have a reliable market for the organic products they plan to produce.

Careful Planning

- In the early part of the transitional period, yields are often depressed and premium prices for certified organic products are unobtainable, so in such cases, use realistic yields and prices when evaluating the feasibility of any project.
- In certain instances, it is preferable to continue using conventional measures early on in the transitional process so as to avoid dramatic yield reduction that jeopardizes the financial well-being of operation.
- Crop growers who are planning to convert their livestock operation should consider certifying their fields first by giving time to learn more about organic livestock management requirements for producing organic feeds.
- As organic certifiers need to see the entire field to be organic, certifiers in such cases will allow new entrants for several years of transition time before certifying the whole farm.
- Parallel production is simultaneous production, processing or handling of organic and nonorganic crops, livestock and other products of similar nature as such type of activity is highly discouraged by certifiers, especially during transition period, but if such parallel production practice is allowed, producers should prepared to deal with significant record-keeping to ensure traceability and organic integrity.

Understanding Soils and Ways of Improvements

- As soil is the heart of the organic farming system, it is important that new entrants should understand the various characteristics and limitations of it on their farm.
- Soil suitability varies from field to field as fields with good drainage, good level of fertility and organic matter, adequate pH, biological health, high legume content, and with less weed and pest pressure are excellent assets that are ready for transition and certification.

- ☆ Certain tools will help in assessing the soil's chemical, physical and biological analyses, soil survey and legume composition field assessments and field yield histories.
- ☆ If crop growers plan to grow crops without raising any livestock, it may be necessary for them to source allowable soil amendments as composted manure, limestone, rock dust, and supplementary sources of nitrogen, phosphorus, potassium, and micro-nutrients.
- ☆ Even with the best of crop rotations which covers green manure crops as legumes (nitrogen-fixing crops), transitional growers will be challenged if they need to get optimal yields without additional livestock manure, compost and/or other off-farm soil inputs.
- ☆ When the inputs are scarce or expensive, producers may benefit from integrating livestock on their farm.
- ☆ In organic production, crop growers must be able to recycle nutrients through proper nutrient management practices by recycling through good manure and compost utilization, crop rotations, cover crops, and by reducing nutrient losses due to leaching, over-fertilization, poor manure and compost management.

Identifying Crops or Livestock for a Particular Situation

- ☆ Before growing crop or raising any livestock, crop grower should consider degree of difficulty to grow or raise the product organically, land and soil suitability, climate suitability, level of demand for product, marketing challenges, capital required, current prices for conventional, transitional and organic products, and profitability over additional workload.

Designing of Good Crop Rotations

- ☆ Once the crops are selected, crop growers should carefully plan the crop rotation(s) and should select the most suitable cover crops.
- ☆ Crop rotations are extremely important management tools in organic farming as it interrupts pest life cycles, suppresses weeds, provides and recycle fertility, and improve soil structure and tilth.
- ☆ Many rotational crops can be cash crops which generate supplemental income.
- ☆ In many farms, land base availability is a limiting factor while planning crop rotations, the transitional plan should include crop rotation strategies.

Identifying Pest Challenges and Methods of Control

- ☆ It is important for crop grower to know about the crop's most common pests, their life cycles, and adequate control measures.

- There are many pests that are significantly important in growing crops such as Colorado potato beetle is for growing potatoes, cucumber beetles is for growing cucurbitaceous crops, flea beetle is for growing many seedlings crops, clipper weevil and Tarnish Plant Bug for growing strawberry crops.
- There are several measures available to reduce pest pressure which covers crop rotation, variety selection, sanitation, floating row covers, catch crops, flamers, the introduction of beneficial insects, bio pesticides, and inorganic pesticides.
- Transitional crop growers should be prepared to use and experiment with certain options, as while considering new type of production, they should discuss pest issues with their agrologists, IPM specialists and/or other existing organic producers to optimize their chances of success.
- Availability of organic supplies has improved significantly over the past few years as new pest control products containing B.T., spinosad, kaolin clay are effective and are currently available to organic growers.
- As seen the types of weeds found on farms evolve with time as crop growers change the way they grow their organic crops and control their weeds.
- By keeping track of weed population, crop growers will be able to refine their crop rotations and improve their control measures.
- Under organic livestock management, cattlemen should provide attentive care to promote health and meets the behavioral needs of various types of livestock.

Conducting Own On-farm Trials

- Successful organic crop growers should continuously try new and/or innovative management practices which should cover cropping, inter-planting, and use of various soils and pest control materials that can be required regularly by organic farmers.

Keeping Good Records

- The record-keeping is an important requirement which will help crop growers to maintain organic integrity.
- Crop growers are expected to keep detailed production, processing and marketing information which should cover information related to every transaction on fields.
- Third party, independent inspectors require crop grower to show detail documentation while inspecting the farm operation.
- Once the record-keeping requirements are understood and the reporting procedure gets established, paperwork will then becomes a normal routine.

Avoiding Common Mistakes

Crop growers should avoid following mistakes:

- Underestimating the need for good transitional and marketing plans.
- Underestimating need to fully understand Organic Standard as organic producers need to understand the standards to know, what is permitted and what is prohibited.
- Transitional crop growers should consider improving their crop rotation, soil and crop management skills, livestock management practices.

1.13.2 Steps Needed to Implement Organic Principles

Principles in Organic Farming

The four principles of organic agriculture are as follows:

Principle of Health

- Organic Agriculture should sustain and enhance health of soil, plant, animal, human and planet as one and indivisible.
- The principle shows that health of individuals and communities cannot be separated from health of ecosystems as healthy soils produce healthy crops that foster health of animals and people.
- Health is wholeness and integrity of living systems and is not the absence of illness, but is the maintenance of physical, mental, social and ecological well-being.
- Immunity, resilience and regeneration are key characteristics of health.
- The role of organic agriculture in farming, processing, distribution or consumption, is to sustain and enhancing health of ecosystems and organisms from smallest in soil to human beings.
- Organic agriculture is intended to produce high quality, nutritious food that contributes to preventive health care and well-being, so it should avoid usage of fertilizers, pesticides, animal drugs and food additives that may have adverse on health effects.

Principle of Ecology

- Organic Agriculture should be based on living ecological systems and cycles, work with them, emulate them and help in sustaining them.
- The principle states that production is to be based on ecological processes, and recycling.
- In this, nourishment and well-being can be obtained through ecology of specific production environment, as in case of crops, it is living soil while

for animals, it is farm ecosystem and for fish and marine organisms is aquatic environment.

- Organic farming, pastoral and wild harvest systems should fit the cycles and ecological balances in nature where cycles are universal but their operation is site-specific.
- Organic management needs to be adapted to local conditions, ecology, culture and scale.
- In this, inputs should be reduced by reuse, recycling and efficient management of materials and energy to maintain and improve environmental quality and in conserving resources.
- Organic agriculture should attain ecological balance by designing farming systems, establishing of habitats and maintenance of genetic and agricultural diversity.
- In this, who produce, process, trade or consume organic products should protect and should benefit common environment like landscapes, climate, habitats, biodiversity, air and water.

Principle of Fairness

- The principle concerns about organic agriculture which needs to build relationships so as to ensure about fairness that regards to common environment and life opportunities.
- Fairness is characterized by equity, respect, justice and stewardship of shared world among people and relationship to other living beings.
- The principle emphasizes that those involved in organic agriculture should conduct human relationships to ensure fairness at all levels and to crop growers, workers, processors, distributors, traders and consumers.
- Organic agriculture needs to be provided to everyone those are involved with good quality of life, and contribute to food sovereignty and reduction of poverty.
- It aims to produce sufficient supply of good quality food and other products.
- The principle insists that animals should be provided with conditions and opportunities of life that accord with their physiology, natural behavior and well-being.
- Natural and environmental resources that are used for production and consumption should be managed in a way that is socially and ecologically just and should be held in trust for future generations.
- Fairness requires systems of production, distribution and trade that are open and equitable and account for real environmental and social costs

Principle of Care

- The principle shows that organic agriculture to be managed in precautionary and responsible manner in order to protect health and well-being of current and future generations and environment.
- Organic agriculture which is a living and dynamic system should responds to internal and external demands and conditions.
- Practitioners of organic agriculture needs to enhance the efficiency and should increase productivity, which should not be at risk of jeopardizing health and well-being.
- New technologies requirement needs to be assessed and existing methods should be reviewed.
- The principle shows that precaution and responsibility are main concerns in management, development and technology choices in organic agriculture.
- Science which is necessary to ensure organic agriculture as healthy, safe and ecologically relates that scientific knowledge alone is not sufficient.
- Practical experience, accumulated wisdom and traditional and indigenous knowledge offer valid solutions, tested by time.
- Organic agriculture needs to be prevented significant risks by adopting required technologies which rejects genetic engineering.
- Decisions should reflect the values and needs of all who gets affected by transparent and participatory processes.

1.13.3 Basic Steps of Organic Farming

The approach of organic farming involves following five principles:

1. Conversion of land from conventional management to organic management
2. Management of the entire surrounding system to ensure biodiversity and sustainability of the system.
3. Crop production with the use of alternative sources of nutrients such as crop rotation, residue management, organic manures and biological inputs.
4. Management of weeds and pests by better management practices, physical and cultural means and by biological control system
5. Maintenance of live stock in tandem with organic concept and make them an integral part of the entire system

Chapter 2

Seed Selection and Treatment under Organic Farming

2.1 Various Seed Treatment Inputs for Organic Farming and their Benefits

2.1.1 Seed

The seed in a plant is the part that develops from the ovules after fertilization. They are enclosed in the fruit which develops from the fertilized ovary. The seeds are formed as a result of sexual reproduction and contains the young embryo which can develop into a new plant.

A seed is a basic part of any plant. The ovules after fertilization, develop into seeds. A seed is made up of a seed coat and an embryo. The embryo is made up of a radicle, an embryonic axis and one (wheat, maize) or two cotyledons (gram and pea). A seed is found inside a fruit which converts into a new plant when we plant it. Hence, the seed is the most important part.

There are hundreds of variations in the seed size, shape, color and surface. The seeds range in size from tiny dust particles, as found in some orchids, to large double-coconuts. The seed surface may be smooth, wrinkled, striate, ribbed, furrowed, reticulate, tuberculation, alveolate, hairy, and pulpy or having patterns like finger prints.

In the seed, life activities are temporarily suspended in order to enable the plant to successfully pass through unfavorable and injurious climatic conditions. On the approach of favorable conditions, the seed resumes active life and grows into full plant. In the form of seeds, a plant can be carried to long distances without special precautions.

2.1.2 Seed Treatment

Seed treatments control the pathogens that occur on or in the seed. The choice of seed treatments may be dictated by whether the pathogen is borne externally or internally. These pathogens may be important for three reasons:

- ☆ Some pathogens do not survive in soil or crop residue and are dependent on the seed borne phase for survival between crops such as fungus that causes loose smut of wheat.
- ☆ If a pathogen can survive in soil or residue, being seed borne may allow it to get a head start and, thus, result in more severe disease such as fungus that causes septoria leaf blotch of wheat.
- ☆ Seed borne pathogens may hitch a ride to new localities in seed shipments.

Methods

- ☆ Due to infection in seeds, wheat crop gets different diseases like: smut, rust, blight *etc.* Due to these diseases, there is adverse effect on annual production of wheat.
- ☆ The infection from such diseases could be minimized or reduced by using only healthy, and quality seeds chosen through hot water seed treatment and treatment with fungicide, which is described in details below:

(A) Hot Water Seed Treatment

- ☆ Initially, soak the seeds in ordinary water for about 4 hours, than dip in hot water 52 °C for 10 minutes or 55 °C for 5 minutes.
- ☆ After that, cool and dry the seeds in shade, and let for fungicide treatment.

(B) Fungicides Treatment of Seed

- ☆ It is compulsory to treat the seeds for producing quality and healthy wheat crop.

- Mix the wheat seeds before sowing with Bevistin or Benlate2 grams (0.2 per cent) per kg seeds in an air tight pot and stir it vigorously for 8-10 minutes.
- Treating the seeds this way helps in controlling diseases and germs on the outer surface of wheat grain such as blast, loose smut, *etc.*

(c) Seed Dressing

- The most common method of seed treatment.
- The seed is either dressed with a dry formulation or wet treated with a slurry or liquid formulation.
- Dressings are applied both on-farm or in specialized seed treatment facilities.

(D) Seed Coating

- A special binder is used with a formulation to enhance adherence to the seed and begin to impact seed size and shape.
- Coatings require advanced treatment application technology.

(E) Seed Pelleting

- The most sophisticated seed treatment technology, resulting in changing the physical shape of a seed to enhance plantability and handling.
- Pelleting requires specialized application machinery and techniques and is the most expensive of the applications.

2.1.3 Seed Treatment Inputs for Organic Farming

Biofertilizers

- They are an essential component of organic farming whose preparations contain live or latent cells of efficient strains of nitrogen fixing, phosphate solubilizing or cellulolytic micro-organisms that are used for application to seed, soil or composting areas to accelerate microbial processes that augment availability of nutrients which can be absorbed by the plants.
- They promote to harvest naturally available biological system of nutrient mobilization.
- The nitrogen fixing biofertilizers are Rhizobium, Azotobacter and Azospirillum.
- Rhizobium is symbiotic nitrogen fixing bacteria used for ground nut, gram, chick pea and soybean with 25g/kg of seed in a region which can fix nitrogen from 57 -100 kg/ha.

- ✰ Azospirillum is associative symbiotic nitrogen fixing bacteria used in seed treatment of rice, sorghum, and maize with 25 g/kg of seed.
- ✰ The inoculation to the seed increases disease resistance and drought tolerance and promotes production of growth promoting substance.
- ✰ Azotobacter is used as living nitrogen fixing biofertilizer for mustard, linseed, pearl millet, finger millets and kodomillets.

Vermicompost

- ✰ Vermicomposting are used as a replacement for farm yard manure.
- ✰ It shows higher yields than farmers practice in maize and onion.

2.2 Methodology for Preparation of Inputs

2.2.1 Biofertilizer Production Technology

- ✰ The requirement of the covered area and type of construction depends on the size of the unit and the total output which can vary from 4500 sq. ft to 10,000 sq. ft.
- ✰ The building should have a main working laboratory, incubation/transfer room, incubation/fermentation room, media preparation room, chemical store, bottling/packaging room, carrier/raw material store and finished goods store.
- ✰ It should have essential equipments, glassware, plastic ware for general laboratory working and quality control.
- ✰ For more volume production, the entire process can be made automatic by installing form-fill-and-seal machine having liquid filling and powder filling units that are installed in the sterile room.
- ✰ During the operation, the machine initially makes a bag where measured amount of carrier is poured with measured volume of broth and bag which are sealed and mixing of contents is done by gently beating the bag using paddles, while moving over the conveyor belt.
- ✰ In this, liquid inoculants can be filled in the bottles using the automatic bottle filling machines under a sterile air flow.
- ✰ As in biofertilizer production, specified bacteria are up-scaled in step-by step processes from slant to bulk fermenters, so the steps of up-scaling should include the test for purity, small conical flasks of 100-250 ml cap, small mother culture fermenters of 1- lit cap, checking purity of portable seed fermenters, large stainless steel fermenters and release for production.
- ✰ In microorganisms for biofertilizers like Rhizobium, Azotobacter, PSB *etc.* that are aerobic and basically surface grower, azospirillum serves as microarophile in N-free conditions which grows aerobically under constant

supply of nitrogen with optimum temperature for different bacteria ranges between 28-35oC.

- In this, rhizobia requires temperature range from 28-30oC, azotobacter and PSB grow requires 30-32oC, while azospirillum requires 34-35oC temperature.
- For mass multiplication of microorganisms, biomass is upgraded serially from slants to big fermenters that involves the preparation of liquid medium in small conical flasks of 100-250 ml capacity with stopper, flasks to have 50 per cent capacity, autoclave at 121oC for 20 min, allowing cooling of medium, transferring flasks and slants of required bacterial culture in a chamber, transferring lapful of bacterial growth from the slants to conical flasks, incubating at appropriate temperatures on rotary shaker for 5-10 days with continuous shaking, checking purity before using culture as the mother culture.
- To serial up-scale, transfer liquid mother inoculum to small seed fermenters or higher capacity conical flasks or aspirator bottle fermenters.
- For inoculation of 10-20 liters of capacity, seed fermenters mother cultures to be prepared in 1-2 liters of capacity aspirator bottle fermenters with outlet tube at base where bottles can be inoculated from their top mouth by opening a plug in the transfer chamber.
- After mother culture is ready in the flasks, cultivate the organisms in small seed fermenters.
- As per the type of organism, quantity of inoculum, rate of aeration and temperature conditions, fermentation period varies from 48 hrs to 7 days.
- With sterile techniques, take small sample to check for purity and for evaluating total cell count by sampling after 24-48 hrs of inoculation for fast growers and 48-72 hrs for slow growers.
- To prepare the bacterial broth in mass scale for commercial production, stainless steel fermenters of 100-2000 lit capacity should be used.
- In this, steel fermenters are steel pressure vessels that are capable of tolerating steam pressure up to 45 psi with provisions for cooling coils, air sprayers, air outlet, inoculation, sampling and drain ports and pot for cleaning and filling *etc.*
- In case of rhizobia, higher capacity fermenters with mechanical stirrers are not very much successful and tend to get contaminated.
- In this, inoculation levels are kept at 0.1 per cent -1.0 per cent which provides 106-107 cells/ml of liquid media with mean generation time to be affected by stage of growth of starter culture and resultant lag phase, temperature of incubation, availability of nutrients, aeration and size of inoculum.

- For seed fermenter, inoculum level of 0.1 per cent - 1.0 per cent is preferred as low inoculum levels extends incubation time and allow sufficient scope for contaminants to grow which allows manufacturer to completely checks for purity, before it is harvested for further use.
- Once microorganisms have been cultured in liquid broth with final population level >5 x 108 to 5 x 109/ml of media, they are mixed with carrier materials, as mixing of broth with carrier material is important stage, as required population level of bacteria cannot be maintained at significantly high level in liquid broth.
- For sterilization of carrier in bags of the required sizes, find the quantity of the carrier to be inoculated with pre-determined quantity of broth to get the final weight.
- To prepare inoculant packet, measure the volume of fully grown broth with total cell count > 5 x 109 cells/ml which needs to be injected aseptically in carrier bags.

2.3 Applying Seed Treatment

The following things need to be considered while treating the seeds:

- Select a non-windy place to treat seeds with fungicide so that the air will not carry away the fungicide.
- Children, aged people, pregnant women are more vulnerable to the poisonous effects of fungicide hence; they should not be involved in treating the seeds.
- Use mask, gloves, and goggles while treating the seeds.
- Wear the clothes that cover the whole body.
- Use only proper or prescribed amount of fungicides.
- Wash thoroughly with water when fungicide touches your skin directly.

Protection of Seeds and Seedlings

- Seeds and seedlings are vulnerable to many soil-borne and foliar pests.
- Insects and pathogens can destroy germinating seeds and young plants, which are relatively tender and lack food reserves to recover from injuries or survive extended periods of stress.
- Under stressful conditions, a number of aggressive or even fairly weak pathogens can become active and cause plant population and yield losses.
- Seed treatments can protect the seed and seedling from attack by certain insects and pathogens. No systemic fungicides or insecticides form a chemical barrier over the surface of the germinating seed which protects

the germinating seed from chewing insects, such as wireworms or soil-borne pathogens, such as pythium.

- Certain systemic seed treatments can protect aboveground parts from sucking insects, such as aphids or foliage diseases, such as rust.
- Systemic fungicides and biological seed treatments can also protect young plants from root rot. Although the duration of protection may be limited, a delay in infection can reduce the losses.
- Seed treatments last only for 10 to 14 days beyond planting, with pesticide breakdown being most rapid under warm, moist conditions.
- Certain active ingredients can protect seedlings considerably longer when applied at the highest labeled rate.

Advantages

- Seed borne pathogens are vulnerable: The seed-borne phase is the weakest link in the life cycle for many plant pathogens. With seed treatments to control seed-borne pathogens is very effective for disease control.
- Precision targeting: Seed treatments are not subject to spray drift as chemicals are applied directly to seeds, little is wasted on no target sites like bare soil.
- Optimum timing: Seeds and seedlings are vulnerable to diseases and insects as compared to mature plants. Applying treatment to seeds will allow pesticides to be present when needed most.
- Low dose: Relatively small amounts of pesticides are used in seed treatments compared to broadcast sprays as it reduces the cost and the potential environmental impact and the probability of chemical residues in harvested grain.
- Easy to apply: Seed treatments are relatively easy and cheap to apply as compared to broadcast sprays.

Disadvantages

- Accidental poisoning: Treated seed looks like food to some animals as hungry livestock birds, such as pheasants or quail consume spilled treated seed.
- Cropping restrictions: Just like other pesticides, some seed treatments may have significant grazing or rotation crop restrictions.
- Limited dose capacity: The amount of pesticide that can be applied is limited by how much will actually stick to the seed. Seed coating technologies are helping to overcome this limitation, but phytotoxicity may still be a problem.

- Limited duration of protection: The duration of protection is short due to a relatively small number of chemicals applied to seed, dilution of the chemical as the plant grows, and breakdown of the chemical.
- Limited shelf life of treated seed: Producing excess treated seed is undesirable as the shelf life of the treated seed is limited. Surplus treated seed cannot be sold for the grain which is a serious limitation for seeds where seed germination and vigor decline relatively quickly.
- Phytotoxicity: Pesticide injury to plant tissues is called phytotoxicity as seed treatments exist in high concentrations on tender tissues of germinating seeds and seedlings having very low phytotoxicity.
- Worker exposure: In course of treating and handling large volumes of seed, workers may be exposed to seed treatment chemicals as aerosols. Inhalation of aerosols and skin contact with seed treatments must be prevented in the seed treatment process.

2.4 Various Characteristics of Seed with their Suitability to Location

- Seed serves as a fundamental input in agriculture and crop growers' first defense against the pest, disease, and other production challenges.
- Seed genetics largely shows the quality and integrity of food, from appearance to flavor to nutritional content.
- Not all organic systems are alike, so climate, soil type and nutrient management vary by farm and region.
- Organic variety trials are important for helping farmers identify new and existing varieties that perform best on-farm and in the region, having specific qualities as demanded by markets.
- Seed variety trials conducted in organic systems helps crop/plant breeders and seed companies to assess the performance of organic varieties and in identifying gaps in organic seed availability.
- The need of certified/quality seeds is assessed on the basis of area sown under different crop varieties, the area covered by hybrid and self-pollinated varieties with seed replacement rate.

2.5 Suitability of Seeds to Organic Farming Practices

- Organic crop producers use organic seeds and planting stocks to protect the integrity of crops.
- Organic growers use conventionally grown seeds when equivalent organic variety is not available, but only if seeds have not been genetically modified or treated with prohibited substances like fungicides.

- ☆ For varieties in organic seed that are available, many crop growers are challenged by lack of sufficient quantity and scarcity of information on a variety of performances under organic conditions.
- ☆ Organic seed availability is low as less investment in such programs is in part due to consolidation in the seed industry, where few regional private seed companies are interested in meeting the diverse needs of organic farmers.
- ☆ Market consolidation also concentrated ownership of plant genetic resources, locking these genetics up with restrictive patents that inhibit access by growers and breeders who wish to protect and expand genetic diversity.
- ☆ Diminished funding for public crop/plant breeding also contributed to the problem.
- ☆ As industry investments in public research increase, research priorities narrow to focus on proprietary and profitable products that benefit larger industries, like agricultural biotechnology.
- ☆ Organic farming works in harmony with nature as organic growers don't use synthetic fertilizers, GMOs or pesticides to fight insects, disease, and weeds and to grow abundant food using other methods.
- ☆ As per seed germination and growth, crop plant needs soil, water, light, and temperature.
- ☆ Seed on the poor ground will not grow very well as seed without water will not grow.
- ☆ A seed without light will only grow a short time while a seed doesn't grow for very long if it is too cold or too warm.
- ☆ Seed growing in soil having decaying plant material or animal manure will grow larger and faster than one grows in soil without the addition of organic matter.

2.6 Demand of Various Varieties in the Market

2.6.1 Introduction

The organic seeds market was valued at USD 1,651.5 million in 2018 and is expected to register a CAGR of 12.7 per cent, during the forecast period (2019-2024). In 2018, North America was the largest geographic segment of the market studied and accounted for a share of around 38.8 per cent of the overall market.

Rising demand for organic products amid augmenting dietary concerns and a change in consumer spending patterns has led to a significant increase in the production of organic crops grown all over the world, which in turn, has enhanced the demand for organic seeds. However, the cost of the seed is becoming a significant economic barrier for farmers in a few countries, which is influencing them to choose conventional seeds over organically produced seeds.

Organic seed consumption has also witnessed growth in recent years, with vegetable seeds garnering the highest share, followed by herbs and ornamentals and others. In the pursuit of healthy food, which is free from chemicals, consumer groups are turning toward outlets that sell pesticide-free' vegetables, which, in turn, is boosting the prices of the organic produce. Hence, most of the farmers are preferably trending toward small-scale organic farming, home gardening, or back yard gardening to satiate the surplus demand.

2.6.2 Market Trends

The organic seeds market is a highly fragmented market, with several small and medium-sized companies witnessing a minimal share in the global market. The higher market shares of players are likely to be attributed to their highly diversified product portfolios and launches of various modern varieties of seeds every year. Moreover, companies are focusing on expanding their business to modern regions, such as Africa and Asia-Pacific, to increase their customer base and brand awareness.

According to the studies conducted by Iko-kai Org, the higher adoption of agricultural chemicals has not only led to the reduction of harmful insects but also beneficial insects. Harmful insects possess the most significant resistance to chemicals when compared to useful insects.

Higher dosages of agricultural chemicals at various frequencies are used to eliminate harmful insects across many parts of the world. This has resulted in the entry of agricultural chemicals into the human body system, resulting in adverse impacts on human health, affecting the autonomic nervous system, endocrine system, sense organs, and other human body parts.

Consumers have become highly vulnerable to the high chemical residue content in the foods. Owing to the increase in awareness and high adoption of strategies to limit agricultural chemicals used, organic farming has gained more importance in recent years.

The rising importance of organic farming and stringent regulations posed by countries to adopt organic farming are encouraging higher sales in the organic seeds market. This is expected to boost the organic seed market's growth potential, globally.

Organic seed is seed produced by the crop that is planted and raised organically for one generation in case of annual crops while two generations for biennial and perennial crops.

It is natural that organic seeds are obtained by purchasing conventional non-treated seeds, grown organically for one season and then sold on to organic farmers as organic seeds.

The organic market is dynamic and evolving rapidly and there is the significant growth potential for organic seed, both at production and in the marketplace.

Common Organic Seeds

- ☆ Basil Genovese (Sweet Italian) Seeds - OG
- ☆ Lettuce Lollo Rossa Seeds - OG
- ☆ Arugula or Rocket/Roquette Seeds - OG
- ☆ Beetroot Bull's Blood Seeds - OG
- ☆ Cherry Tomato Yellow Pear Seeds - OG
- ☆ Chilly Jalapeño Pepper Seeds - OG
- ☆ Arugula or Rocket/Roquette - Microgreens Seed Pack - OG
- ☆ Basil Sweet Italian Microgreens Seed Pack - OG
- ☆ Cabbage Golden Acre Seeds - OG
- ☆ Cabbage Red Acre Seeds - OG
- ☆ Cherry Tomato Large Red Seeds - OG
- ☆ Cucumber Marketmore Seeds - OG
- ☆ Lady Finger Red (Bhindi) Seeds - OG
- ☆ Lettuce Buttercrunch Seeds - OG
- ☆ Parsley Plain Leaf Seeds - OG
- ☆ Radish Cherry Belle Seeds - OG
- ☆ Radish Early Scarlet Globe Seeds - OG
- ☆ Radish Pink Beauty Seeds - OG
- ☆ Swiss Chard Five Color Silverbeet Seeds - OG
- ☆ Swiss Chard Rainbow Seeds - OG
- ☆ Tomatillo Green Seeds - OG
- ☆ Basil Thai Seeds
- ☆ Kale Blue Curled - Microgreens Seed Pack - OG
- ☆ Zucchini Black Beauty Seeds - OG

2.6.3 Organic Seeds Market: Segmentation

- ☆ Global Organic Seeds market can be segmented based on type, distribution channel, farm type, and region.
- ☆ Based on type, the market of organic seeds is segmented into vegetable seeds, field crop seeds, fruits and nuts seeds and other vegetation seeds.
- ☆ Based on the distribution channel, the organic seeds market is segmented into a direct and indirect segment where an indirect segment can be sub-segmented into modern trade, specialty stores, online retail, and others.

- On the basis of farm type, organic seeds market is segmented into indoor and outdoor where indoor segment is further sub-segmented into vertical farming, greenhouse, and hydroponics and outdoor segment is sub-segmented into the field, gardens and nurseries.

2.6.4 Organic Seeds Market: Dynamics

- Increasing demand for healthier food products and beverages is universal which allows opportunities for players in the global organic seeds market.
- The ever-rising demand for food products with standard ingredients is expected in significantly high sales of organic products.
- Players operating in the organic seeds market are expected to leverage the opportunity during the forecast period.
- With the increasing recurrence of clean label product, launches with organic labels by major food and beverage companies impelled a niche market of organic food products into the mainstream market which is anticipated to cause the growth of global organic seeds market over the forecast period.
- The inadequate infrastructure of regulatory bodies for organic farming concerning to coexistence of different agricultural practices *i.e.* organic, conventional and genetically engineered yields contaminated crops which restrain the growth of organic seeds market in the near future.

2.6.5 Organic Seeds Market: Regional Outlook

- On the basis of regional outlook, organic seeds market is segmented into different regions.
- On analyzing the organic seeds market at a global level, Europe and APAC dominate the organic seeds market owing to heightening demand of organic products that couples with high shelf visibility of organic products in modern retail formats like the supermarket/hypermarket.
- The market region is anticipated to exhibit a considerable growth rate owing to increasing import demand of organic products from developed markets and improving living standards of the middle class of emerging and developing markets, which enable consumers to opt for healthier and more natural food products.

2.7 Characteristics of Seed with their Suitability to Location

2.7.1 Characteristics of Quality Seed

Quality Seed must be genetically pure having:

- Breeder/nucleus: 100 per cent
- Foundation seeds: 99.5 per cent

- Certified seeds varieties: 98 per cent
- Certified hybrid seeds:95 per cent
- Certified hybrid cotton: 90 per cent
- Certified hybrid castor: 85 per cent

Quality Seed should have required level of physical purity such as:

- For all crops: 98 per cent
- Carrot: 95 per cent
- Ragi: 97 per cent

Quality Seeds should be free from other crop seeds such as designated inseparable crop seeds that are plant seeds cultivated crops found in seed fields where seeds are similar to crop seed that is difficult to separate by mechanical means, causing physical admixtures with crop seeds only when these crop mature at same time when seed crop matures. It covers barley: oats and wheat seeds.

Quality Seeds should be free from objectionable weed seeds, as seeds of weed species that are harmful in one or more of following ways;

- Size and shape of weed seeds are similar to crop seed and difficult to separate them by mechanical means
- Weeds growth habit is determinate type and compete with crop for all resources
- Weed plant parts are poisonous or injurious to human and animals
- Weed plants serves as alternative hosts for pests and diseases
- *Ex.* Bersem: chicory
- Paddy: wild paddy (*Oryza sativa* var. *fatua*)
- Cucurbits: Wild cucurbits spp.
- Lettuce: wild lettuce
- Bhendi: wild Abelmoscous spp
- Wheat: *Convolvulus arvensis* (Hirankuri)

Quality Seed should be free from designated diseases, such as diseases that are specified for seed certification causing contamination of seed lot

- *Ex*; loose smut of wheat (*Ustilago tritici)*
- Kernal smut of sorghum (*Sphacelotheca sorghi*)
- Grain smut of pearl millet (*Tolyposporium penicillariae*)
- Ergot of bajra (*Claviceps microcephala*)
- Early blight of tomato (*Alternaria solani*)
- Little leaf of brinjal (Datura virus-2) *etc.*

Quality Seed should have high germination and vigor

Quality Seed should have optimum moisture content such as in:

- Cereals: 10-12 per cent
- Pulses: 7-9 per cent
- Oilseeds:6-7 per cent
- Vegetables: 5-6 per cent

2.7.2 Organic Seeds as per Location

Lower Altitudes and Mid-hilly Areas

- The region is the best with rich resources of soil, agro climate, agro-biodiversity, forest biodiversity, wetlands, rainfall, flora and fauna.
- In this, cereals form an overwhelming proportion of 85 per cent of total gross cropped area while rice covers about 70 per cent of total cultivated area.
- Apart from maize and millets, other important crops that grow in the region include potato, oilseeds, cotton, jute, pulses, chilies, sugarcane, sweet potato, *etc.*
- In this, fruits and vegetables occupy about 5–6 per cent of cultivated area.
- In the mid-hilly areas, food grain, cash crop and horticultural crops includes ginger, turmeric, arecanut, pineapple, orange, litchis, large cardamom, passion fruit, papaya, banana, guava, black pepper, cashew nuts, jack fruit, *etc.*
- In high hills, common fruits that grows include plums, pears, peaches, apricots, apples, *etc.*
- The common grown vegetables in the area includes potato, cabbage, cauliflower, radish, carrots, beans, broccoli, *etc.*

Agro-climatic Zones

- In such areas, among the field crops, paddy is dominant along with potato, maize and millet.
- Varieties of vegetables like cabbage, carrot, cauliflower, broccoli, beet root, radish, *etc.*, grows here.
- In such areas, common fruits such as apple, plum, peaches, pears, walnut, mango, lemon, guava, banana, mandarin orange, pineapple, banana, papaya, litchi *etc.* are grown.

2.8 Suitability of Seeds to Organic Farming Practices

2.8.1 Introduction

- Organic seed is produced by non-chemical farming or is not treated with chemicals.
- Being organic, however, whether applied to seed or crop produce, takes much wider and deeper meaning that cuts across disciplines and dimensions.

2.8.2 Opting Organic Seed

Organic crop growers, prefer organic seeds as it:

- Promote Biodiversity: Over the last century, thousands of local varieties of seed disappeared with estimated losses of around 75 per cent of agricultural genetic diversity.
- Keep Collective Resources: Seed sovereignty equals food freedom as rampant consolidation swept by the seed industry which results in much of the world's seed resources that are controlled by certain corporations.
- Avoid Genetic Engineering: Certified organic seeds are free of content from genetic engineering.
- Promote Biodiversity: Seed crops are longer than food crops where it results in more applications of chemical herbicides and pesticides where allowance for chemical pesticides and herbicides applied to seed is much higher as the seed is considered as a non-food crop.
- Protect pollinators' health: Organic seed is not pre-coated with dangerous neonicotinoid pesticides. In this, neonics are systemic where the pesticide is drawn in plant's vascular system from its roots and is expressed throughout the plant in pollen and nectar which pollinators forage for and drink.
- Encourage continued organic research: Organic seed is better suited for low-input organic systems which regionalized research that leads to adaptation of varieties and makes them more robust. Seed farmers can select varieties of plants that perform well under local ecological conditions and pressures such as pests and disease, and changing climate.
- Celebrate varieties selected for quality: Organic researchers and farmer-breeders consider flavor and nutrition in their breeding selections, instead of focusing on traits like high yield and extended shelf life.
- Preserve agrarian tradition: The future of organic farming depends on sharing the skills and knowledge of agriculture which includes seed stewardship.

- ☆ Support family farmers: Organic seed farming is intensive and often family-scale which substitutes management skill for toxic chemicals that leads to rural economic development in hiring local labor.
- ☆ Contribute to a thriving organic industry: The development of vibrant organic seed trade results in seed systems that are suited to ecological, economic, and localized challenges and the needs of organic farming at every scale.

2.9 Resistance of Varieties to Pests and Diseases

2.9.1 Introduction

In the growth of agriculture, improved varieties of crops play an important role in development, so the main objective of crop science research involves in development of improved varieties in different crops. The new improved varieties of crops are developed by Crop Research Institutes or State Agricultural Universities or by Private Seed Companies.

The crop varieties are tested for a minimum period of three years, before consideration of release for cultivation. The important characters for which new varieties are evaluated/tested include:

- ☆ Yield
- ☆ Resistance to disease and pests
- ☆ Quality of product
- ☆ Resistance to adverse environmental factors
- ☆ Adaptation to different agro-climatic conditions

2.9.2 Variety Release

Norms for Release of New Variety

There are certain norms that should be considered for the release of the new variety. Besides distinctness, uniformity, and stability, the new variety should be superior to previously released variety in one or more of the following characteristics:

- ☆ Yield potential of grain, oil, fiber, fodder, vegetable or other economic product.
- ☆ Resistance to biotic and abiotic stresses.
- ☆ Quality of oil, fiber, protein, fodder, vegetable, grain, *etc.*
- ☆ Maturity duration.
- ☆ Suitability for general cultivation over a wide range of environmental conditions.
- ☆ Suitability for the machine harvesting, *etc.*

Types of Varieties

Pure-Line Variety

- It is developed in self-pollinated crops by pure-line selection.
- The progeny of a single homozygous plant obtained by selfing is called pure-line.
- The variety is released from the single best progeny.
- The base population, which is used for the development of pure-line variety is always a heterogeneous population.
- A pure-line variety is a homozygous and homogeneous population.

Mass Selected Variety

- It is developed in self and cross-pollinated species.
- In this, the mass selection is more common in cross-pollinated crops than in self-pollinated crops.
- The variety is released from the single best progeny.
- In cross-pollinated species, a mass-selected variety is a mixture of several homos and heterozygotes that can be heterozygous and heterogeneous population.
- A mass-selected variety has a broad genetic base and wide adaptation.

Clonal Variety

- It is developed in asexually propagated crops like potato, sugarcane, sweet potato, *etc.*
- The progeny of single best clone is released as a cultivar.
- Such varieties are heterozygous but homogeneous, as all plants of a clone have exactly the same genotype.
- Clonal cultivars have wide adaptation than pure-line cultivars.
- A heterogeneous base population is used for the development of clonal variety.

Synthetic Variety

- A synthetic variety is developed in cross-pollinated crops such as maize.
- For the development of synthetic variety, good general combining inbreds are identified.
- In this, inbreds with good general combining ability are intermated in all possible combinations and hybrid seed is obtained.
- The hybrid seed of all combinations is mixed together in equal quantity to constitute a synthetic variety.

- In this, the synthetic variety has several heterozygotes which is maintained by open pollination to allow some degree of selfing.
- In cross-pollinated species, a mass-selected variety is a mixture of several homos and heterozygotes that can be heterozygous and heterogeneous population.
- The yield level of synthetic variety is always higher than open-pollinated variety but lower than the yield of single and double-cross hybrids.

Composite Variety

- It is developed in cross-pollinated species.
- It is developed by mixing the seed of various open-pollinated genotypes which are similar in height, maturity, seed color, seed size, *etc.*
- The composite variety has heterozygous and heterogeneous population.
- It consists of many homozygotes and heterozygotes.
- The varieties are highly adaptable to environmental fluctuations.

Multiline Variety

- It is developed in self-pollinated crops.
- A multiline variety is a mixture of several isogenic lines, related lines or unrelated lines.
- In this, each component line has a different gene for disease resistance.
- Such varieties are homozygous but heterogeneous and have very good buffering capacity due to the broad genetic base.

Mutant Variety

- A mutant variety is developed through induced mutagenesis.
- Mutant varieties are developed in self-pollinated and asexually propagated species and rarely in cross-pollinated species.
- A mutant variety differs from parent variety in one or two characters.
- Such varieties are developed to induce disease resistance, earliness, dwarfness and achieve improvement in quality.

Transgenic Variety

- A variety that is developed by the technique of genetic engineering is called transgenic variety.
- Such varieties are developed to solve specific and difficult problems to be solved by conventional methods of plant breeding.
- Transgenic varieties with herbicide resistance, insect resistance and improved quality have been developed in several crop plants.

- ☆ Such varieties can be developed in all three groups of crops *viz.* Self-pollinated, cross-pollinated and asexually propagated.

Pedigree, Bulk, and Backcross Bred Varieties

- ☆ Pedigree and bulk bred varieties are developed in self-pollinated species.
- ☆ These varieties are developed from the progeny of a single best homozygote.
- ☆ homozygous and homogeneous Varieties have a narrow genetic base.
- ☆ The backcross method is used both in self and cross-pollinated species for improvement of oligogenic character like disease resistance.
- ☆ Backcross bred variety is similar to parent variety except for the gene which has been transferred into parent variety.

Hybrid Variety

- ☆ The F1 population used for commercial cultivation is a hybrid variety.
- ☆ Hybrid varieties are developed in the cross and self-pollinated species.
- ☆ Hybrid varieties are more common in cross-pollinated species than in self-pollinated species.
- ☆ Hybrids are of different types, *viz.* Intervarietal hybrids and interspecific hybrids.
- ☆ Hybrids are heterozygous but homogeneous.
- ☆ For hybrid varieties, fresh seed has to be produced every year.

Features

- ☆ Variety release is the main stage informal seed chain that links plant breeding to farmers.
- ☆ It covers all procedures and actions that take place from the time when the breeder decides to enter a promising new line/variety for official testing till the time when a new variety is made available for multiplication as the first step on road to general cultivation by farmers.
- ☆ Release relates to physical activities linked with multiplication program or promotional campaign.
- ☆ In public-sector varieties, registration is followed directly by release, provided sufficient early generation material to initiate multiplication.
- ☆ In this, public breeders are judged on their output of varieties and are reluctant to use limited resources for maintenance and seed multiplication.
- ☆ The new varieties should be marketed once registered.
- ☆ For registering, tests for distinctiveness, uniformity, stability, and value for cultivation and use needs to be conducted first before registration.

- The registration establishes legal ownership of the new variety.
- The seed laws on variety testing and release variable and inconsistent.
- The variability and inconsistency of seed laws make it costly for seed companies to release and market new varieties.
- A new variety must be tested each time it is to be marketed in respective countries, even if it is developed for sale across a wide range of agro-ecologies.
- In many cases, the variety was unreleased based on its merit and uniqueness.

2.10 Availability and Procurement of Off-farm Inputs as Required

- Agricultural inputs are products allowed for use in organic farming which include feedstuffs, fertilizers, and crop/plant protection products, cleaning agents and additives for food production.
- The farm inputs procurement allows crop growers to purchase inputs with chemicals directly from manufacturers at significantly more minimum prices.

While procuring farm inputs, crop growers show an understanding of:

- Transparent pricing: For procuring farm inputs, crop growers should see there should not be any bundling or other hidden fees, or complicated rebates while purchasing inputs.
- Standard pricing: For procuring farm inputs, crop growers should see there should not be any tech fees, one-off discounts or regional penalties.
- Best price guarantee: For procuring farm inputs, crop growers should see that while making a purchase, the manufacturer/distributor will guarantee the lowest price offered for purchasing and delivery.
- Flexible fulfillment: For procuring farm inputs, crop growers should see there should be the option of selecting between direct delivery and self-pickup which will help crop growers in saving enough money.
- Time-saving and simplicity: For procuring farm inputs, crop growers should see that while procuring the input, there will not be more time involvement and input should be made available without any hassle.

Chapter 3

Soil Nutrient Management under Organic Farming

3.1 Authorized Soil Testing Centers

3.1.1 Introduction

- ☆ Soil testing in respective regions is authorized by respective state/ government which serves as the best method available for determining how much nutrient is needed for optimum production of most crops.
- ☆ For certain crops, use of plant analysis serves as preferred method of determining crop nutrient needs.
- ☆ Making nutrient-application decisions without soil testing or plant analysis can be a risky endeavor because this will probably result in reduced profitability from either under or over-applying nutrients; in addition, over-application of nutrients can result in reduced water quality.
- ☆ It is important to follow recommended procedures for taking soil and plant samples.

3.1.2 Facilities/Equipments in Laboratories

Soil sampling laboratories have following physical facilities:

a. Soil preparation room

b. Sample storage space

c. Main laboratory area

d. Research room

e. Chemical storage room
f. General office and hall

Soil sampling laboratories have following equipments and instruments:

1. Conductivity bridge
2. Shaking apparatus
3. pH meter
4. Rincing and stirring
5. Photoelectric colorimeter
6. Flame photometer
7. Centrifuge
8. Analytical balance.
9. Triple beam balance
10. Multiple dispensing equipment
11. Washing assembly
12. Pipetting assembly
13. Calculating machine
14. Constant holtage stabilizer
15. Water distillery
16. Oven
17. Air compressor

Materials and Equipments required in soil sampling laboratories include:

a. Samples sheets with result.
b. Electrodes for PH meter.
c. Glass atomizer for flame photometer.
d. Conductivity Cell
e. Lamps – for colourimeter
f. Filter for colourimeter
g. Standard solutions for phosphorous, potassium and organic carbon.
h. Glasswares
i. Chemicals
j. Weighing scoop
k. Campe hairbrushes
l. pH meter
m. Colourimeter

n. Flame photometer
o. Conductivity bridge
p. Vacuum pump
q. Platining kit
r. Shaking machine

3.1.3 Functions

The functions of authorized soil testing laboratories is to:

- ☆ Analyze soil samples that are collected by farmers or from farmers by Agricultural Officers.
- ☆ Analyze irrigation water samples for EC, pH, cations, and anions to assess the quality based on different parameters and suggesting suitable ameliorative measures for different soil conditions and crops.
- ☆ Rate the soil test value for soil samples collected during the particular year as low, medium and high.
- ☆ Conduct trials related to soil fertility to solve the site-specific problems.

3.2 Soil Test Reports and Analyzing Nutrient Needs

3.2.1 Soil Sampling for Testing

For testing of soil sample of a particular region, organic crop grower has to collect the soil sample from different areas of the field. Soil sampling is done for checking soil pH, phosphorus, potassium, and micronutrients which should be done after every three to four years as these values do not change rapidly even with fertilization. Soil samples for soil nitrate should be taken 1-3 months before planting each wheat crop. In most dryland or rainfed wheat production areas, the wheat root will have a depth of 5-foot with a full profile of soil water.

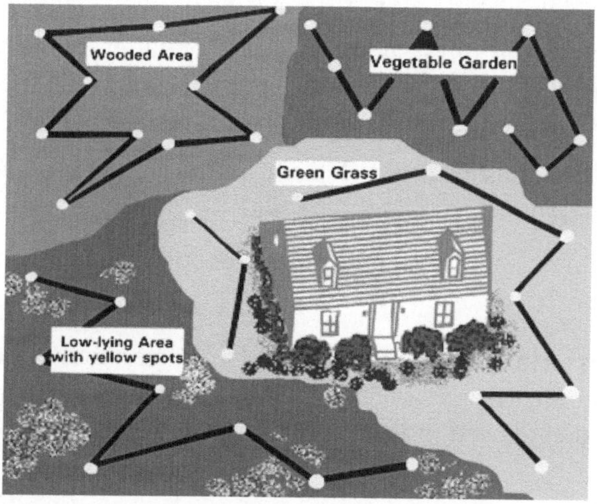

Effective and profitable fertilizer recommendations require an accurate sample of the field. It is very important that the soil sample should accurately represent the field to be fertilized. Previous crop history, previous lime and fertilizer rates, and soil characteristics will cause variance in results if not accounted while sampling.

- Soil sampling is most challengeable work, as few grams of soil sample represents for a given area, hence soil samples needs to be taken in such a manner that the collected sample should reflect the true fertility of soil for any part of the land area.
- If the field is levelled and soil appears to be uniform, only one composite sample if taken properly could be enough for an area of 4-5 ha.
- Variation in slope, colour, texture, crop growth, unusual spots and management practice should be taken into account and separate sets of composite samples must be collected from each such area.
- Recently fertilized plots, bunds, channels, marshy tracts and area near trees, wells, cattle dung and compost piles or other non-representative locations must be avoided during the sampling.

3.2.2 Soil Test

Soil testing is used for quick characterization of fertility status of soils which gives farmers a service leading to better and economic usage of fertilizers for better soil management practices to increase agricultural production. Soil test shows an analysis of soil sample which determines nutrient content, composition, and other characteristics like acidity or pH level. A soil test determines fertility, or expected growth potential of the soil which shows nutrient deficiencies, potential toxicities from excessive fertility and inhibitions from the presence of non-essential trace minerals.

Soil testing is performed by commercial labs that offer a variety of tests, targeting groups of compounds and minerals. It involves collecting soil samples, preparation for analysis, chemical or physical analysis, interpretation of analysis results, and finally making fertilizer and lime recommendations for crops. The laboratory tests often check for plant nutrients in three categories such as:

- Major nutrients: nitrogen (N), phosphorus (P), and potassium (K)
- Secondary nutrients: sulfur, calcium, magnesium
- Minor nutrients: iron, manganese, copper, zinc, boron, molybdenum, chlorine

Soil testing is an important diagnostic tool that determines the nutrient needs of plants and environmental assessments. It facilitates fertilizer composition and dosage selection for land employed in both agricultural and horticultural industries. Soil testing can also be used to identify application rates of waste materials containing nutrients or other elements that could harm the environment.

Soil testing is the best method available for determining how much nutrient is needed for optimum production of most crops. For certain crops, the use of plant analysis serves as the preferred method of determining crop nutrient needs. Making nutrient-application decisions without soil testing or plant analysis can be a risky endeavor because this will probably result in reduced profitability from either under or over-applying nutrients; in addition, over-application of nutrients can result in reduced water quality. It is important to follow the recommended procedures for taking soil and plant samples.

3.2.3 Soil Test Report

- ✰ Soil test report from the lab provides a representation of soil sample results in which information about different levels of various nutrients is described.
- ✰ In this, optimal and excessive levels mean that the nutrient concentration in the soil is more than adequate for optimum crop growth.
- ✰ The data obtained from soil analysis would be meaningless until they are correlated with crop yield.
- ✰ On the basis of soil testing results, the recommended fertilizer dose for the crops is modified for making fertilizer recommendations to the farmers.

Format

Once the soil test report gets prepared, it is the duty of a wheat cultivator to collect the report and understand the basic information for cultivating the wheat crops. In order to understand the report, the wheat cultivator should study each and every column of the report.

The format of soil test report includes:

1. *Organic Matter*

Percent organic matter is a measurement of the amount of plant and animal residue in the soil. The color of the soil is usually closely related to its organic matter content, with darker soils being higher in organic matter

2. *Phosphorus*

Shows three types of phosphorus tests; P1 (weak Bray), P2 (strong Bray) and Bicarbonate P (sodium bicarbonate).

3. *Potassium*

It measures the available potassium in the sample where its optimum level varies with crop, yield, soil type, soil physical condition, and other soil related factors.

SOIL ANALYSIS REPORT

E INFORMATION SHEET: 123456

LAB NUMBER	SAMPLE IDENTIFICATION	ORGANIC MATTER	PHOSPHORUS			NEUTRAL AMMONIUM ACETATE (EXCHANGEABLE)				pH		CATION EXCHANGE CAPACITY	PERCENT BASE SATURATION (COMPUTED)				
		MODIFIED LOI	P_1 (WEAK BRAY)	P_2 (STRONG BRAY)	BICARBONATE P	POTASSIUM K	MAGNESIUM Mg	CALCIUM Ca	SODIUM Na	SOIL pH 1:1	BUFFER INDEX	C.E.C.	% K	% Mg	% Ca	% H	% Na
		percent RATE	ppm RATE	ppm RATE	ppm RATE	ppm RATE	ppm RATE	ppm RATE	ppm RATE			meq/100g					
F LAB NUMBER	G SAMPLE IDENTIFICATION	(1)		(2)		(3)	(4)		(5)	(6)	(7)	(8)			(9)		

LAB NUMBER	NITRATE-N										SULFUR S ICAP	ZINC Zn DTPA	MANGANESE Mn DTPA	IRON Fe DTPA	COPPER Cu DTPA	BORON B SORB. DTPA	EXCESS LIME RATE	SOLUBLE SALTS 1:1	
	SURFACE			SUBSOIL 1			SUBSOIL 2			Total lbs/A									
	ppm	lbs/A	depth (in)	ppm	lbs/A	depth (in)	ppm	lbs/A	depth (in)		ppm RATE	ppm RATE	ppm RATE	ppm RATE	ppm RATE	ppm RATE		mmhos/cm RATE	
			(10)								(11)	(12)	(13)	(14)	(15)	(16)	(17)	(18)	

4. Magnesium and Calcium

It shows the levels of calcium and magnesium in the soil that are affected mainly by soil type, drainage, liming and cropping practices.

5. Sodium

It shows the sodium level in soil sample

6. Soil pH

This shows soil pH measures, active soil acidity or alkalinity in soil sample.

7. Buffer Index

It shows an index value for determining the amount of lime to apply on acid soils with pH less than 7.0.

8. Cation Exchange Capacity

It measures the soils ability to hold nutrients like potassium, magnesium and calcium as well as other positively charged ions as sodium and hydrogen in soil sample.

9. Percent Base Saturation

It shows part of Cation Exchange Capacity occupied by given cation in soil sample.

10. Nitrate-Nitrogen

The soil test measures nitrate-nitrogen (NO3-N) which is water soluble and readily available for plant uptake.

11. Sulfur

The soil test measures several forms of sulfur that can be readily available.

12. Zinc

It shows extraction of the zinc in soil sample.

13. Manganese

It shows detail about manganese extraction reading using DTPA extraction process.

14. Iron

It shows details about iron extraction using the DTPA extraction process in the sample.

15. Copper

It shows detail about copper extraction using DTPA extraction process in the sample.

16. Boron

It shows reading of boron extraction from the soil using DTPA/Sorbitol where adequate levels shows from 1 to 1.5 ppm.

17. Excess Lime Rate

It shows detail about visual rating of free lime where soil with high amounts of free lime have problems with availability of major and minor elements to the plant.

3.3 Types of Inputs Needed as per Quantity at different Crop Stages

3.3.1 Crop Stages

Organic crop growers should need knowing about different growth stages of the plant which will help in determining the nutrients and inputs which includes:

First Stage – Seed Germination

- ☆ The plant's life cycle starts with the germination of seed with the use of water, oxygen, and temperature for the seed to germinate.
- ☆ Many seeds are unaffected by light or darkness, but some seeds will not germinate until they see sufficient light.
- ☆ The correct temperature required water was important for the seed to germinate as a dry seed needs water.
- ☆ Many annual vegetables have optimal germination temperatures between 75-89 F (24-32 C) while certain species as radishes or spinach germinates at a low temperature of around 40 F.
- ☆ When seed drinks water, it expands while enzymes and food supplies in seed become hydrated.
- ☆ The seed becomes activated and starts to go to work and provides energy for the growth process.
- ☆ The growth of the root takes place which shows plant growth using the ideal environment.

Second Stage – Vegetative and Growth

- ☆ The root system supports plant growth which leads to the beginning of the vegetative stage.
- ☆ Plant nutrient requirements call for a large amount of nitrogen for the production of chlorophyll.
- ☆ Growth during this stage is dedicated to stem, branch and leaf areas.
- ☆ The substantial growth over the life cycle of the plant occurs in the vegetative stage which will continue unless hindered by a change in environment or lack of water and nutrients.

Third Stage – Reproductive, Flowering, and Fruit

- In the reproductive stage, the plant's energies are directed to the production of flowers, fruits, and seeds.
- In this, nitrogen along with phosphorus and potassium serves as the main nutrients.

3.3.2 Organic Inputs at Crop Stages

- Organic inputs are derived from the processing of plant and animal products that the farmer brings to his crop in order for it to express its production potential. The most commonly accepted organic inputs are manure, slurry and green compost.
- By decomposing and fermenting, organic materials from the farm or outside allow organic inputs to form. Plants' inputs are derived from plant decomposition. They include compost, pruning residues, organic preparations (infusion, decoction), ramial chipped wood, *etc.*
- Organic inputs such as various types of composts, biofertilizers, *etc.* are not only important in organic farming but are also of prime importance in integrated agriculture with balanced use of fertilizers for the sustenance of soil fertility.
- The main role of organic inputs is to be used either as fertilizer for crops or as a soil amendment. Fertilizers are used when the objective is to provide nutrients directly to the plant via the soil or by spraying whereas amendment is used when the objective is to improve the soil's physical, chemical, and biological properties.
- Mass adoption of vermicomposting technology and the use of neem seed kernel sprays by farmers is indicative of the usefulness of such strategy.
- To take advantage of the growing awareness of organic agriculture, various types of organic and biological inputs have been launched and are being sold to farmers. Some of such products are the results of research and are being promoted by state agencies also, but some of the products have been launched without much scientific understanding and their quality and usefulness are questionable.
- In this, some organic fertilizers and biofertilizers have been covered under Fertilizer Control Order and their standards and quality control parameters have been defined.
- Biofertilizers or microbial inoculants are carrier-based ready to use live bacterial or fungal formulations, which on application to plants, soil or composting pits, help in mobilization of various nutrients by their biological activity.

- ☆ For increased availability of nitrogen and phosphorus, nitrogen-fixing biofertilizer and phosphate solubilizing biofertilizer together are used in equal quantities in pulses and legume oilseeds such as moong, urad, lentil, pea bengal gram, arhar, cluster bean, groundnut, soybean, berseem, leucern and in nonlegume crops like wheat, rice, maize, bajra, oats, barley, mustard, sesame, niger, onion, potato, sugarcane, cotton, *etc.*

3.4 Methods of Application, Timing, and Doses of Different Inputs

3.4.1 Introduction

- ☆ Once the soil testing is done, the organic grower needs to have information about the timing, dosage and application of organic inputs for related crops in terms of nutrient requirement and nutrient already present in the soil.
- ☆ The paddy crop growers should use 63 Kg of urea, 53kg of DAP, 51kg of single super phosphate with 34 kg of potash for one acre of paddy crop to provide 39 kg of nitrogen, 33 kg of phosphorus and 21 kg of potash.
- ☆ The maize crop growers should use 83kg of urea, 59 kg of DAP, 35 kg of single super phosphate and 38kg of potash per acre with 49 kg of nitrogen, 32 kg of phosphorus and 23 kg of potash.

3.4.2 Application of Recommended Doses of Inputs

- ☆ The adoption of recommended doses benefits the growers in terms of improvement in the yield, net returns and better soil health.
- ☆ The use of recommended doses of fertilizers/inputs is effective in mobilizing macro and micronutrients in soil and in improving soil fertility.
- ☆ The adoption of recommended doses fertilizers increases crop yield in respective crops.
- ☆ Fertilizer/inputs usage is crucial for sustainable intensification and for raising farm productivity under increasing land constraints and declining soil fertility.
- ☆ Fertilizers for the particular crop and field depend on information obtained from a soil test report, whose recommendations will be given on soil health cards.
- ☆ The application of recommended fertilizers/inputs is influenced by factors such as awareness of crop growers about advantages in applying recommended doses of fertilizers, availability of fertilizers, convenience, sources of purchase of fertilizers, weather conditions and time of fertilizer application, *etc.*

- There are nowadays many crop growers who show a willingness to adopt practices so as to maintain better soil health for better yields.
- Mostly, medium and large growers adopted recommended doses of inputs/fertilizers in their crop while maximum adaptability was found in large and small organic growers and minimum adaptability was found in marginal organic growers.
- The required dosage by soil test organic growers apply 85 kg of urea, 58 kg of DAP, 25 kg of potash and 67 kg of complex fertilizers per acre. The required dosage by soil test organic growers apply 85 kg of urea, 58 kg of DAP, 25 kg of potash and 67 kg of complex fertilizers per acre.

3.4.3 Method of Application

- Different kinds of organic input/manures like farmyard manure, vermicompost/biogas waste, bio-fertilizer, green manure and manures like city compost were used by crop sample growers in cultivation practices.
- The efficiency of fertilizer utilization by crops is mainly influenced by the method of application of fertilizers.
- The application of fertilizer causes many changes in the soil such as chemical changes which can positively or negatively influence its productiveness.

3.4.4 Timing of Inputs

Soil type affects the timing and frequency of fertilizer application.

Nitrogen Application

- Nitrogen needs careful management, as it is very susceptible to be lost from soils.
- It can be lost from soil by leaching, denitrification, erosion and surface volatilization.
- It is more readily leached in sandy soils than in fine-textured soils.
- If nitrogen is properly unapplied, its loss can account for up to 50-60 per cent of the applied amount.
- If nitrogen is applied too early before the plant requires, then a significant part of it gets lost before the crop takes it up, so time nitrogen is in the soil before the plant takes it up which should be minimized.
- For proper nitrogen timings, split the nitrogen application which helps in reducing the risk of nitrogen loss and improves the efficiency of the application.

Phosphorous Application

- ✰ Movement of phosphorus in soil is very slow, so roots can uptake phosphorus only from close surroundings.
- ✰ When phosphorous is initially added to soil with fertilizers, it gets available insoluble and available form.
- ✰ Phosphorous quickly becomes unavailable for plants in a process known as fixation.
- ✰ As phosphorus applied remains at the topsoil layer, its losses will be through surface runoff and soil erosion, so the timing of phosphorus application and frequency of applications is important.
- ✰ If phosphorous is applied before rain or heavy, then irrigation will lead to loss through runoff and erosion, so splitting of phosphorus application is highly recommended.

3.5 Organic Soil Enhancement

3.5.1 Introduction

- ✰ Healthy soils are a complex web of life, teeming with earthworms, beneficial fungi, and bacteria which smell good and are moist and crumbly.
- ✰ Roots are able to penetrate easily, deep into the soil, so plants growing in healthy soils have fewer pest and disease problems.
- ✰ To enhance soil means to improve it using additional materials.
- ✰ The soil amendments or soil conditioners improve the physical nature of the soil and reduce compaction, aerating the soil to allow water and nutrients to move through it and reach plant roots.
- ✰ Soil enhancement also adds nutrients to the soil, helps retain moisture and maintains the correct pH balance.

3.5.2 Common Soil Enhancement include

Garden Soil

- ✰ It is soil enhancement for in-ground use rather than containers or raised beds usage.
- ✰ For this, a grower has to work it in existing soil to improve aeration and moisture retention.
- ✰ The soil includes fertilizer to feed the plants.
- ✰ Sphagnum peat moss absorbs water, slowly releasing it for use by plant roots and lightens clay soil, providing aeration and add mass to sandy soil, preventing leaching of nutrients.

Composted Manure

- It is an odorless farm byproduct.
- To improve aeration and moisture retention, it enriches the soil.
- Dehydrated manure is a similar product which contains less moisture.

Mushroom Compost

- It is a mixture of straw, peat moss, and other organic components, formulated for use in commercial mushroom production.
- The mixture is used for one round of growing and is packaged as enhancement for fields.
- Topsoil is intended for repairs of the garden or lawns which fill holes, levels the ground, correcting eroded areas, *etc.*
- Topsoil should be unused for planting or potting unless it blends with other enhancements like composted manure, vermiculite, perlite, peat moss, and aged wood bark.

The soil enhancement also covers:

- Lime - raises soil pH, reducing acidity
- Sulfur - lowers soil pH, increasing acidity
- Gypsum - improves aeration of compacted soil, helping it drain more efficiently
- Perlite - improves aeration and drainage
- Vermiculite - improves moisture retention and aeration

3.6 Production Methodologies for different Types of Composts and Other Inputs

3.6.1 Production Methodologies for Compost

- Composting is the art of producing rich, sweet-smelling decomposed products rather than wet, smelly, fly attracting, rotting mess.
- Good compost contains a range of ingredients, wider the range, more likely the full range of nutrients.
- Good compost includes cow, horse, and poultry manure, comfrey and arrowroot leaves, legumes, wood ash, grass clippings, weeds, seaweed, kitchen scraps, leaves, hair, sawdust, feathers, soaked newspaper, animal bedding, fish waste, and crop residues nutrients.

3.6.2 Methods of Composting

Coimbatore Method

In this

- Composting is done in pits of different sizes as per the availability of waste material.
- The layer of waste materials is initially laid in the pit and is moistened with the suspension of 5-10 kg cow dung in 2.5 to 5.0 I of water and 0.5 to 1.0 kg fine bone meal which sprinkles uniformly.
- Layers are laid one over the other till the material rises 0.75 m above ground level and is finally plastered with wet mud and left undisturbed for 8 to 10 weeks.
- Plaster is removed, and the material gets moistened with water, given turning and made in a rectangular heap under shade which is left undisturbed till use.
- There appears anaerobic decomposition followed by aerobic fermentation which is reversed in the case of Bangalore method.

Indore Method

In this

- Organic wastes are spread in cattle sheds which serve as bedding.
- Material with urine-soaked is removed every day which forms in a layer of about 15 cm thickness.
- Urine-soaked earth gets scraped from cattle sheds mixed with water and sprinkled over layer of wastes two or three times in a day.
- Layering process continues till fortnight where a thin layer of decomposed compost is sprinkled over top and heap shows turning and reformed.
- Old compost will act as inoculum for decomposing the material.
- The heap is left undisturbed for a month which is thoroughly moistened and gives a turning.
- Compost is ready for application next month.

Bangalore Method

In this

- Dry waste material of 25 cm thick is spread in the pit and thick suspension of cow dung in water is sprinkled over for moistening.
- The thin layer of dry waste is laid over the moistened layer.

- The pit is filled with dry layers of material, and cow dung suspension is done till it rises 0.5 m above the ground level which is left exposed without covering for 15 days.
- The pit is given a turning and is plastered with wet mud and left undisturbed for about 5 months or till required.
- Thorough decomposition is not as compared to Indore compost or as much as Coimbatore compost.

3.7 Proper Application Process and Schedule

3.7.1 Timings Application

- The timing of inputs/fertilizer applications has a significant effect on crop yields.
- Proper timing of input/fertilizer application increases yields, reduces nutrient losses, increases nutrient use efficiency and prevents damage to the environment.
- Applying inputs/fertilizers at wrong timing will lead to nutrient losses, waste of fertilizer and damage of crops.
- The optimum timing for fertilizer application is determined by the nutrient uptake pattern of organic crops.

Timings of Inputs

Soil type affects the timing and frequency of fertilizer application.

Nitrogen Application

- Nitrogen needs careful management, as it is very susceptible to being lost from soils.
- It can be lost from soil by leaching, denitrification, erosion and surface volatilization.
- It is more readily leached in sandy soils than in fine-textured soils.
- If nitrogen is properly unapplied, its loss can account for up to 50-60 per cent of the applied amount.
- If nitrogen is applied too early before the plant requires, then a significant part of it gets lost before the crop takes it up, so the time nitrogen is in the soil before the plant takes it up, should be minimized.
- For proper nitrogen timings, split the nitrogen application which helps in reducing the risk of nitrogen loss and improves the efficiency of the application.

Phosphorous Application

- Movement of phosphorus in soil is very slow, so roots can uptake phosphorus only from close surroundings.
- When phosphorous is initially added to soil with fertilizers, it gets available insoluble and available form.
- Phosphorous quickly becomes unavailable for plants in a process known as fixation.
- As phosphorus applied remains at the topsoil layer, its losses will be through surface runoff and soil erosion, so the timing of phosphorus application and frequency of applications is important.
- If phosphorous is applied before rain or heavy, then irrigation will lead to loss through runoff and erosion, so splitting of phosphorus application is highly recommended.

3.8 Various Soil Activating Inputs (Jeevamrut, Farmyard Manure *etc.*)

3.8.1 Jeevamrut

- It is among the four pillars of Zero budget natural farming which is prepared from cow dung, cow urine, pulse flour, jaggery, and soil.
- It improves the quality and yield of fruits, vegetables, and other crops.
- It contains an enormous number of microbial loads which multiplies and act as soil tonic.
- Its application enhances microbial activity in soil and ensures availability and uptake of nutrients by crops.
- It promotes immense biological activity in the soil and enhances nutrient availability to crop.
- It is low-cost improvised preparation that enriches the soil with indigenous microorganisms, so required for mineralization.
- During 48 hours of the fermentation process, aerobic and anaerobic bacteria presents in cow dung and urine which multiply as they eat organic ingredients.
- In this, undisturbed soil is added to preparation, as inoculate of native species of microbes and organisms.
- It helps to prevent fungal and bacterial plant diseases.
- It is only required for the first 3 years of transition, after which, the system becomes self-sustaining.
- It is applied to the crops twice in a month in irrigation water or as 10 per cent foliar spray.

Preparations

200 liters of water in a barrel

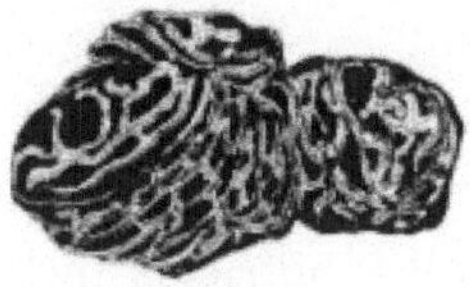

10 Kg fresh local cow dung

2 Kg of pulse flour

2 Kg of Jaggery

5-10 liters aged cow urine

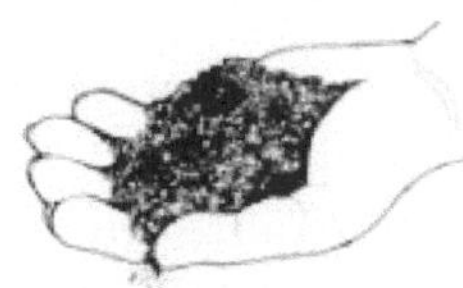

handful of soil

- ☆ Put 200 liters of water in a barrel
- ☆ Add 10 Kg fresh local cow dung and 5-10 liters aged cow urine
- ☆ Add 2 Kg of Jaggery (brown sugar), 2 Kg of pulse flour and handful of soil from bund of farm

- ☆ Stir the solution well and allow it to ferment for 48 hours in shade.

Application

- ☆ For liquid foliar spray, use 5 per cent -10 per cent in water while for soil, use 100-200 lts per acre during irrigation.
- ☆ As per the growth of crop, grower should use it once at an interval of 7–15 days.
- ☆ Spraying during dawn or dusk on any crop will promote the growth, flowering and increase in yield.

3.8.2 Farmyard Manure

- Farm manure is a valuable soil improver that enhances and restores a range of natural properties of the soil.
- Farmyard manure is a fertilizer composed of waste products, typically dung and urine, produced by farm animals
- The waste products contain a wealth of nutrients, including the extremely important element that plants need–nitrogen.
- It contains 5-6 kg of nitrogen, 1.2-2.0 kg of phosphorus and 5-6 kg of potash per tonne.
- The nutrient supplying and improving capacities of manure in sandy soils are limited to 2-4 vegetative periods due to their rapid decomposition and mineralization in soil.

Nutrients Present

Nitrogen	0.5000 per cent
Phosphorus	0.2500 per cent
Potassium	0.4000 per cent
Calcium	0.0800 per cent
Sulfur	0.0200 per cent
Zinc	0.0040 per cent
Copper	0.0003 per cent
Manganese	0.0070 per cent
Iron	0.4500 per cent

Features

- Increases soil fertility
- A natural source of available nitrogen
- Adds humus and slow-releasing nutrients to the soil
- Aids water and nutrient retention
- Helps break down heavy soils
- Adds structure to light and sandy soils
- Attracts worms to the soil
- Ideal for mulching

3.9 Preparation Methodology for Soil Activating Agents

The soil-activating agent includes soil pH, potassium carbonate, potassium hydroxide with alginic acid extract and marine alga organic matter.

Soil Activation by Sterilization

- Soil sterilization is known to alter physicochemical characteristics of soil, plant growth and community structure of the newly developed bacterial population.
- Soil sterilization also increased the root length, surface area, volume and number of tips.
- Soil sterilization is frequently used to remove microbial activity pertaining to microbial inoculations, soil enzymes, and degradation/sorption/mobility of pesticides and other xenobiotic.

Materials

- Method of soil sterilization includes heat treatments (dry heat and autoclaving), fumigation (formaldehyde, propylene oxide, chloroform and methyl bromide) and high-level dakua-irradiation.
- For this, arable soil layer of 0–20 cm with N, 0.07 per cent; organic matter, 0.85 per cent; maximum water-holding capacity, 30.5 per cent; CaCO3, 1.8 per cent; pH (1:1, soil water), 7.2; EC (1:1, soil water), 672 $\mu S\ m^{-1}$; CEC, 9.6 meq g–100; sand 63.0 per cent; silt, 20.7 per cent and clay, 16.3 per cent.
- The soil should be air dried, sieved (<2 mm) and stored at room temperature.
- Before autoclaving, soil needs to be moistened to around 45 per cent WHC which should be conditioned at room temperature for 48 hours.
- The soil which is of 2-cm thick layer in glass dishes needs to be autoclaved at 121°C for one hour.
- For dakua-irradiation, 5-kg portions of air-dried soil were packed in cardboard boxes which should be lined with polyethylene sheet and irradiated at 50 kGy using 60Co commercial gamma irradiator.

Process

- To activate soil by sterilizing harmful microorganisms in early-stage, microorganisms useful for planting is left to remain in soil.
- The soil-activating agent is obtained by adding 0.5-1 mass per cent of sodium hydrogen carbonate and 0.01-0.05 mass per cent of citric acid to an aqueous solution of chlorine dioxide in a concentration of 500-700 ppm.
- The soil-activating agent is sprinkled over the soil in a ratio of 2,000-2,500 liters per Tan (about 992 m2), and microorganisms in soil can be selectively sterilized and soil for planting can be activated in an early stage.

3.10 Planning for Green Manuring and Biomass Recycling Methods and Strategies

3.10.1 Green Manures

- Green manures form an important part of organic systems, but many growers use them more effectively.
- It is used regularly in crop rotation for the shorter period between cash crops to maintain soil N concentrations and gives long-term leys.
- Green manuring is the practice of incorporating plant material into the soil from a crop grown for the purpose of soil improvement.
- There are three different types of green manure as fixers, lifters, and retainers.
- Fixers are legumes that fix nitrogen from atmosphere in roots.
- Lifters are plants that use available nutrients in soil and holds them in their leaf and stem structure.
- Retainers are quick-growing plants that soak the soil N and prevent leaching and losses.
- Fixers produce nitrogen ready to be released in the soil in 2–3 weeks after being incorporated.
- Fixers include species such as red and white clover, vetches, crimson clover, lupins, Lucerne and grain legumes.
- Lifters include crops like mustard, phacelia, and ryegrass.
- Retainers include covers like forage rye, volunteers and even weeds.
- Green manures are usually, but not necessarily incorporated in-situ and generally at the onset of flowering.

Green manuring has certain advantages to the grower such as:

- Adding organic matter to the soil
- Increasing biological activity
- Improving soil structure
- Reduction of erosion
- Increasing the supply of nutrients available to plants
- Reducing leaching losses
- Suppressing weeds
- Reducing pest and disease problems
- Providing supplementary animal forage
- Drying and warming the soil

Planning for Green Manuring

- ☆ Green manures are legumes that are grown without the intent to harvest but to improve soil fertility.
- ☆ They are an important part of organic grain rotation as they supply nitrogen needed to fuel the crops in the rotation.
- ☆ It is significant to consider how and when green manures will be grown when designing crop rotation.
- ☆ It is grown every two to five years in the rotation, as per the farm, soils, and climate.
- ☆ There are many ways to integrate green manures in crop rotation, with options that will fit most any Prairie organic grain production system.
- ☆ With rising nitrogen fertilizer prices and ever-increasing requirements to farm in an environmentally sustainable way, green manures are fast becoming viable ways to cut input costs, add fertility and improve the soil.
- ☆ There is a wide variety of green manures to choose from including clovers, medicks, mustards, and grasses.
- ☆ Legumes are very popular as they fix nitrogen, but other species offer benefits like improved soil structure and weed suppression.
- ☆ Green manures can be grown for widely differing periods to suit particular needs and varies from six weeks, for short break in intensive vegetable rotation too many years as grass ley.
- ☆ Green manures have many different benefits but no one species will offer all of these.
- ☆ Grown as a single species, or in mixtures, the right choice of green manure depends on aims and circumstances of each individual farmer or grower.
- ☆ They can be used for a whole range of reasons as their benefits are very diverse.
- ☆ When choosing which to grow, each farmer needs to analyze his/her specific aims and circumstances and use these to help make the right decision which includes soil type, farming system, previous cropping, future cropping plans, and climate.

For the planning of green manure, the grower has to work out on:

Crop Rotation Plan

- ☆ If a grower needs to avoid building up soil pests and diseases, it's important to follow a crop rotation plan.
- ☆ Many simple crop rotation plans like 3-course rotation work on Roots, Legumes, Brassica takes place, so grower needs to take this into account when choosing green manure.

- ☆ There are many green manures that don't need crop rotation plan as Buckwheat, Forage Rye and Phacelia, which are used mainly to add organic matter to the soil.

Soil Preparation and Matching to the Green Manure

- ☆ Regardless of soil type, the soil is going to dig over and free from weeds, so grower needs to know about what type of soil is present.
- ☆ Certain green manures are used in specific soils only and if grower wants to get the maximum nitrogen/soil benefit then he/she needs to match the green manure with the right soil.

Time of Year to Grow

- ☆ Green manure is thought of as autumn/winter crop due to its ability to assist in reducing the impact of soil erosion from winter rains and allowing roots to grow deeper in the soil to assist with aeration and breaking up heavy soils such as clay.

Sowing Process

Sowing green manure requires following steps:

- ☆ Prepare the soil by digging it over and removing any weeds, break up any clumps, rake the seedbed evenly to fine tilth and lightly tread the soil
- ☆ For small green manure seeds, broadcast sow the green manure which ensures grower to get good coverage for weed suppression.
- ☆ Larger green manure seeds like Field Beans and Forage Peas can be sown in furrows 10cm apart in rows 20cm apart to a depth of 5cm
- ☆ Rake over the seedbed and water well
- ☆ Keep the seedbed just moist until cover crop germinates and becomes established.

3.10.2 Biomass Recycling Methods and Strategies

Bioreactor System

- ☆ In biomass separator, the flow from the bioreactor is passed which has been used to mimic closely.
- ☆ In this, biomass in a vessel having porous walls by which clear liquid flows, leaving behind the biomass inside.
- ☆ In this, wastage flow is taken directly from the vessel.
- ☆ The tangential flow membrane filtration is used as biomass separator, having high recirculation gets removed from bioreactor which passes over

the filter and returned to the bioreactor, but a small percentage will pass through the filter for discharge the biomass-free effluent.

☆ With such a system, the flow velocity of the effluent fluid is less than the settling velocity of biofloc, which discharges clarified effluent and retains biomass in a reactor that takes wastage directly from the mixed bioreactor.

☆ In Garrett configuration, as shown, the biomass is wasted from concentrated solids recycle stream.

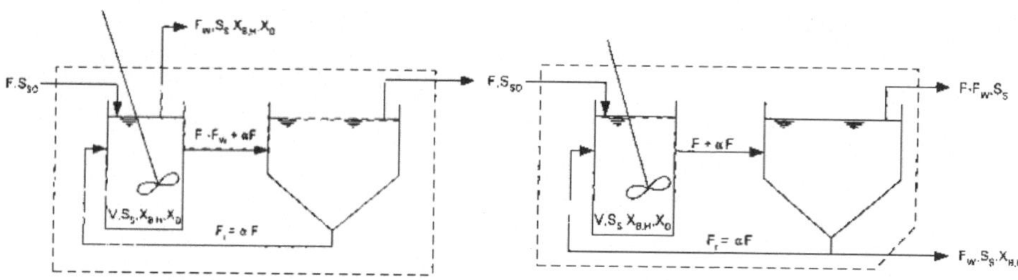

☆ As the bioreactor is completely mixed, concentration of constituents in the effluent stream leaving it and flowing to the settler are the same as in the case of bioreactor.

☆ In this, if the settler is operated properly, solids blanket will be small so that the mass of solids in the settler is small as compared to the mass in bioreactor.

☆ With this, there results less reaction in the settler so that the concentrations of soluble constituents in recycle stream are the same as those in bioreactor.

☆ In this, as the settler is perfect, all the biomass entering it gets returned to the bioreactor where recycle stream has no impact on the biomass concentration in the bioreactor.

☆ Here, the impact of the recycle flow rate on the performance of the bioreactor with conventional configuration shows that the recycle flow rate has on waste solids concentration whose impact on performance of bioreactor with conventional configuration is an understanding of the effect that the recycle flow rate has on waste solids concentration.

☆ It is shown that the concentration of constituent in wastage flow will be the same as concentration in solids recycle stream as wastage flow comes from it.

3.11 Calculating Nutrient Needs Based on Soil Test Reports

3.11.1 Soil Test

☆ provides present quantitative information on nutrient content and nutrient-supplying capacity of soil by measuring the per cent base saturation, that shows the ratio of base ions on exchange sites, quantitative measurement of cation exchange capacity, measurement of soil's potential to hold and exchange cation nutrients and parts per million of nutrients such as N, P, K, Mg, *etc.*

- allow for periodic monitoring of soil chemical properties to maintain soil nutrient levels in established optimal ranges and serves as an accurate indicator of nutrient depletion or accumulation.

3.11.2 Soil Analyses

- provide an accurate determination of soil's textural classification that helps a grower to anticipate how soil will respond to cultivation, soil's nutrient and water holding capacity.
- provide quantitative data allowing for comparison of given soil's nutrient and chemical profile with established benchmarks for each property to identify nutrient levels that are above or below optimal benchmarks which serve as remedied over the long term with annual soil amending.
- helps to identify nutrients that exist at very low or very high levels resulting in acute plant nutrient deficiencies or toxicity, so that soil nutrient imbalances can address by amending and/or supplemental fertilizing program

3.11.3 Soil Analysis Properties

Unbalanced plant-nutrient status results from the use of organic or inorganic fertilizers. Many organic materials do not contain nutrients in balanced amounts. Many nutrients from animal sources have an excess of phosphorus and potassium relative to plant demand for nitrogen. In soil, nutrients can accumulate to levels exceeding crop needs with repeated application based on plant nitrogen needs. In the case of organic materials, regularly soil tests to monitor phosphorus, potassium, and salt accumulation. Nutrient amendments need to be tested regularly, as similar organic materials vary considerably in nutrient content as per the source, handling and conditions present when the plant was living. Organic crop growers should check and analyze the following soil properties and should take decisions.

Percentage of Organic Matter

- The measurement of percent organic matter content of soil like raw and soil organic matter.
- As climate influences, the ability of organic matter to accumulate, no benchmarks for soil organic matter can be applied.
- Soil organic matter higher than 2 per cent, and ideally 3–5 per cent organic matter content in cool temperate climates is desirable.

- Measuring changes in soil, organic matter over time is useful in monitoring the impacts of the soil management program.

Estimated Nitrogen Release

- The amount of plant available nitrogen in lbs/acre which are estimated to be released by mineralization of presently existing soil organic matter during the single growing season
- The amount of nitrogen liberated from the decomposition of organic matter is dependent on soil biological activity which is influenced by soil and environmental conditions like soil temperature and other climatic conditions, soil pH, the chemical composition of organic matter, soil aeration and soil moisture, so ENR figure shown in soil report should be strictly estimated by grower
- ENR data are a set of figures that are used in calculating the nitrogen budget for the crop(s).
- The grower should use 60 per cent of ENR values as listed in soil tests for annual crop production.
- The grower should use 80 per cent of ENR value as listed on soil tests for perennial crop production.

Extractable Phosphorus

- There are two types of tests that are commonly used to show phosphorus availability at different pH levels.
- Available phosphorus (P1) Weak Bray method for acidic soils (pH<6.2) where optimum levels of readily available phosphorus (P1) should be 30–40 ppm
- Olsen sodium bicarbonate extraction for basic soils (pH>7.5) for certain states soil tests, where optimum levels of phosphorus should be 40–60 ppm with pH of 6.0 or more for cold season vegetable crops or 15–25 ppm for warm-season vegetable crops

Extractable Potassium/Potash (K)

- It is the amount of exchangeable of potassium which is found in a given soil sample.
- The optimal levels of available potassium in a given soil are 2–5 per cent of base saturation
- The total amount to have this is based on CEC of soil where optimal levels for coarse-textured soils range from 150–175 ppm while for heavy textured soils, it will be 175–250 ppm

Extractable Magnesium (Mg)

- It is the amount of exchangeable magnesium in ppm which is found in a given soil sample.
- The optimal levels of available magnesium where the total amount to have this depend on CEC of soil
- The optimal range of magnesium in a given soil is 10–20 per cent of base saturation.
- Soils with magnesium levels over 23 per cent of base saturation exhibit drainage problems and should need attention.

Extractable Calcium (Ca)

- It is the amount of exchangeable calcium in ppm found which is in a given soil sample.
- The optimal level of calcium in a given soil is 65–75 per cent of base saturation where the total amount for this is based on Cation Exchange Capacity of soil

Sodium (Na)

- It is the amount of exchangeable sodium in ppm found in given soil sample.
- The optimal levels of sodium Na in a given soil are 0–5 per cent of base saturation while the total amount for this depends on the Cation Exchange Capacity of soil and should be less than 100 ppm.
- Sodium levels above 5 per cent of base saturation cause serious soil structure issues that affect water movement, aggregation, and tilth.

pH

- It is the measurement of acidity or alkalinity of a given soil.
- For pure water pH of 7.0 is neutral while for lemon juice, pH is less.
- A high pH is referred to as basic or alkaline (pH ~9.0).
- The optimal pH range should be around 6.3–6.8 for mineral soil, 5.5–6.0 for organic soil
- For adjusting pH, the grower should use calcium carbonate, calcium/magnesium carbonate, calcium sulfate, elemental sulfur.

Buffer Index

- It is an index that is based on soil pH that is used to estimate the amount of time which is required to raise a soil having a pH of 6.5 or less to several higher pH levels. (6.0 and 6.5)

Hydrogen

- ☆ To measure hydrogen ion concentration in meq/100g of soil in a given soil sample.
- ☆ Since hydrogen ion concentration increases, soil acidity also increase which shows a decrease in pH.

Cation Exchange Capacity (CEC)

- ☆ It is a measurement of soil's ability to hold and exchange cation nutrients like Ca, Mg, Na, K, hydrogen with growing plants.
- ☆ Cation Exchange Capacity of soil is influenced by the amount and types of clays and organic matter in soils.
- ☆ Soils with higher clay and organic matter content have higher Cation Exchange Capacity and are most fertile.

Base Saturation

- ☆ It is the relative percentages of major cation nutrients that occupy cation exchange sites in a given soil.
- ☆ To balance, the crop grower should use nutrients in well-balanced state which should cover potassium (K) 2–5 per cent, magnesium (Mg) 10–15 per cent, calcium (Ca) 65–75 per cent, hydrogen (H) 0–20 per cent, sodium (Na) 0–5 per cent,

Nitrate Nitrogen (NO_3^-)

- ☆ It is the amount of water-soluble nitrogen (nitrate) in ppm found in a given soil sample.
- ☆ Nitrate with ammonium is a plant-available form of nitrogen that results from nitrogen fixation and mineralization actions of soil microbes.

Sulfate (SO_4^-) Sulfur (S)

- ☆ It is the total amount of sulfur in ppm which is found in given soil sample.
- ☆ The optimal levels of sulfur S are 25–35 ppm

Micronutrients/Trace Elements

- ☆ Zinc (ZN): It is the extractable amount of zinc in ppm found in a given soil sample having optimal levels as 1.1–3.0 ppm (DTPA extraction)
- ☆ Manganese (Mn): It is the extractable amount of Mn in ppm found in a given soil sample having optimal levels as 9–12 ppm (DTPA extraction)
- ☆ Iron (Fe): It is the extractable amount of Fe in ppm found in a given soil sample having optimal levels as 11–16 ppm (DTPA extraction)

- ☆ Copper (Cu): It is the extractable amount of Cu in ppm found in a given soil sample having optimal levels as 0.9–1.2 ppm (DTPA extraction)
- ☆ Boron (B): It is the extractable amount of B in ppm found in a given soil sample having optimal levels as 0.6–1.2 ppm (hot water extraction)

Excess Lime

- ☆ It is visual observation and rating of carbonates in a soil sample.
- ☆ The high levels of free lime available may interfere with nutrient availability.

Soluble Salts

- ☆ It is the total measurement of soluble salts by electrical conductivity (ECE).
- ☆ The high levels show a higher risk of plant toxicity due to salt accumulation from fertilizers, poor irrigation water, or chemical contamination.
- ☆ The optimal levels of available soluble salts will be less than 2.0 mmhos/cm or < 100 ppm.

Chloride

- ☆ It is available in a small amount in all soils.
- ☆ It is supplied by irrigation water and organic sources like manure and compost.

3.12 Planning for the Need of Off-Farm Inputs and their Procurement

3.12.1 Need

- ☆ Small farmers often do not own farm equipment beyond the basic moldboard plow, a scotch cart, and a harrow.
- ☆ With the advent of CA, farmers need having suitable equipment for the cropping system. The success of this largely depends on the resource endowment of the farmers and income sources.
- ☆ The use of intensive inputs in agriculture and access to plentiful fossil energy has enabled an increase in food production and standard of living, providing food and livelihood possibilities where they may have been previously limited, unavailable or untenable.
- ☆ With the Green Revolution, their result in increasing usage of high yield seeds, fertilizers, and pesticides with diesel and electricity, so agricultural inputs take significant share in the production system of modern agriculture
- ☆ An important factor that drives the usage of aggregate agricultural input is the amount of area devoted to cropland.

- ☆ Fossil fuel is the main source of energy to cultivate, harvest and transport crops and is vital in the production and manufacturing of inputs such as farm machinery, pesticides, and fertilizers.

Procurement

- ☆ The production and use of safe feed will enhance plant performance and improves profitability, so it is the duty of crop growers to obtain safe ingredients, as it is impossible to produce safe feed without safe ingredients.
- ☆ Before procuring, feed ingredients should be monitored and inspected by the crop grower.
- ☆ Feed ingredients should meet acceptable and statutory standards for levels of pathogens, mycotoxins, pesticides, and undesirable substances to give raise to consumers' health hazards.
- ☆ If feed ingredients are produced on the farm, then crop growers should take care that it should meet the requirements as established for feed ingredients that sourced off the farm.
- ☆ The purchase of inorganic fertilizer should be made from a reliable source in order to have guaranteed content of plant nutrients and the absence of chemical contamination like heavy metals and fluorine.

Chapter 4

Weed Control under Organic Farming

4.1 Various Types of Weed

4.1.1 Introduction

- Weed is a plant considered undesirable in a particular situation.
- It includes plants unwanted in human-controlled settings like farm fields, gardens, lawns, and parks.
- It has no botanical significance as a plant that is a weed in one context is not a weed when growing in a situation where it is in fact wanted, and where one species of plant is a valuable crop plant, another species in the same genus might be a serious weed, such as a wild bramble growing among cultivated loganberries.
- Volunteer crops are regarded as weeds in a subsequent crop.
- Many plants that people widely regard as weeds on top are intentionally grown in gardens and other cultivated settings, in which case they are sometimes called beneficial weeds.
- Weed applies to any plant that grows or reproduces aggressively or is invasive outside its native habitat.

Features

- A number of weeds like the dandelion and lamb's quarter are edible with leaves or roots used for food or herbal medicine.
- It prevents pest insects from securing a crop as their presence disrupts the incidence of positive cues that pests use to discover their food.

- ✰ Weeds act as a living mulch and provide ground cover to reduce moisture loss and in preventing erosion.
- ✰ It improves soil fertility by bringing up nutrients like calcium and nitrogen from deep in soil with taproot and clover hosts nitrogen-fixing bacteria in its roots, fertilizing the soil directly.
- ✰ The dandelion species breaks up the hardpan in overly cultivated fields and help crops to grow deeper root systems.
- ✰ Many garden flowers originate as weeds in cultivated fields and have selectively bred for garden-worthy flowers or foliage.

4.1.2 Types of Weeds

There are three types of common weed plants as per their growing characteristics which include:

Annual Types

- ✰ It germinates and spread by seed having average lifespan of less than a year.
- ✰ It includes winter and summer types.
- ✰ It grows from seed, develop into mature plant, set flowers and seeds and die after seeds gets shed.
- ✰ They maintain their populations by production of seeds with each generation.
- ✰ Winter annuals like chickweed, germinate in late summer/early fall, go dormant in winter and actively grow during spring.
- ✰ Summer annuals like lambs quarters, germinate in spring, grow throughout the summer and goes with the arrival of cold weather.

Biennial Types

- ☆ It completes their life cycle in two years, germinates and forms rosettes in first year and produce flowers and seeds in second year.
- ☆ These weeds are the best controlled in young growing stages of first-year plant.
- ☆ It includes bull thistle and garlic mustard.

Perennial Types

- ☆ These plants are able to live for two years or more.
- ☆ It returns every year and produces long tap roots with seeds.
- ☆ It includes dandelions, plantain and purple loosestrife.
- ☆ These plants are the most difficult to control.

4.2 Use of Different Methods to Control Weeds such as Plastic Mulch

4.2.1 Preventative Weed Control

- It refers to the control method that aims to prevent weeds from being established in a cultivated crop, a pasture or a greenhouse.
- Examples include the use of the certified weed-free seed, transporting weed-free hay, making sure farm equipment is cleaned before moving from one location to another, and screening irrigation water to prevent weed seeds from traveling along irrigation ditches.

Preventative practices of weed include:

- Cleaning tillage and harvesting equipment of weed seeds and vegetative structures
- Planting certified
- Weed-free crop seed
- Controlling weeds in barnyards, around structures and along fencerows, roadways and ditch banks

4.2.2 Techniques of Weed Control

Controlling weeds in crop production involves a wide range of techniques which includes:

Cultural

- It refers to a technique that involves maintaining field conditions such that weeds are less likely to become established and/or increase in number.
- Examples of cultural weed control would be crop rotation, avoiding overgrazing of pastures or rangeland, using well-adapted competitive forage species, and maintaining good soil fertility.

Mechanical

- It refers to a technique that involves the use of farm equipment to control weeds.
- Mechanical/physical techniques destroy weeds or make the environment less favorable for seed germination and weed survival.
- Such techniques include hand-pulling, hoeing, mowing, plowing, disking, cultivating and digging.
- Mulch can be used to control weeds and helps in keeping the soil moist, prevent frost damage and adds to the organic matter in the soil.

- ☆ Mulch can be organic or can be of plastic sheeting, where plastic is removed at the end of the season and is disposed of in an environmentally friendly manner.

Biological

- ☆ It refers to a technique that involves the use of hereditary enemies of weed plants to control the germination of weed seeds or the spread of established plants.
- ☆ It is rapidly expanding area of weed control with many examples which includes sheep to control tansy ragwort or leafy spurge, cinnabar moth and the tansy flea beetle to control tansy ragwort, the chrysolira beetle to control St. John's Wort, and the use of goats to control brush on rangeland.

Chemical

- ☆ It refers to a technique that involves the application of a chemical (herbicide) to weeds or soil to control the germination or growth of the weed species.
- ☆ It includes chemicals to control weeds which include 2,4-DB; EPTC; bromoxynil; and paraquat.
- ☆ Herbicides are crop protection chemicals used to kill weedy plants or interrupt normal plant growth.
- ☆ Herbicides provide a convenient, economical and effective way to help manage weeds and allow fields to be planted with less tillage showing earlier planting dates and provide extra time to do other tasks that farm/personal life require.
- ☆ With reduced tillage, soil erosion reduces which reduces soil from entering waterways and decreases the quality of surface water.
- ☆ Without herbicide use, no-till agriculture becomes impossible, but the use of herbicide also carries environmental, ecological and human health risks.
- ☆ Herbicides may not be a necessity on some farms/landscape settings, but without the use of chemical weed control, mechanical and cultural control methods become more important.
- ☆ Many kinds of herbicides can be selected based on many factors that determine when, where and how particular herbicide can be used most effectively.

4.2.3 Method to Control Weeds by Plastic Mulch Method

Mulching

- ☆ Mulching is the process of covering topsoil with plant material like leaves, grass and crop residues.

- Mulch cover enhances the activity of soil organisms like earthworms and helps to create soil structure with many small and large pores by which rainwater can infiltrate easily in soil and reduces surface runoff.
- The type of material used for mulching mostly influences its effect, so the material selected should be easily decomposed to protect soil and should provide nutrients to crops while decomposing.
- The germination and nourishment of many weeds can be controlled by mulch treatment.
- The mulches can cover the soil surface or work as a physical barrier and prevent the germination of weeds or physically control seedling appearance.
- The less weed intensity is recorded in polythene and straw mulch plots compared to the chemical and unmulched.

Mulching is mainly employed for:

- Moisture conservation in rainfed areas
- Reduction of irrigation frequency and water saving in irrigated areas
- Soil temperature moderation in greenhouse cultivation
- Soil solarisation for control of soil-borne diseases
- Reduce the rain impact, prevent soil erosion and maintain soil structure.
- In places where high-value crops only to be cultivated

Sources of Mulching Material

- Cover crops, grass, and weeds
- Crop residues/straw
- Pruning material from trees and hedges
- Wastes from agricultural processing or forestry

Application of Mulch

- Mulch should be applied before or at the onset of the rainy season when the soil is most vulnerable.
- If the layer of mulch is not too thick, seeds or seedlings can be directly sown or planted in between the mulching materials.
- On vegetable plots, it is best to apply mulch only after young plants have become hardier, as they may be harmed by decomposition products from fresh mulch material.
- If mulch is applied before sowing or planting, the mulch layer should not be too thick in order to allow seedlings to penetrate.
- Mulch can be applied in established crops and is best down directly after digging the soil.

- It can be applied between the rows, directly around single plants or evenly spread on the field.

Plastic Mulch

- Mulching is done either by the use of biodegradable mulching materials or by various mulch films.
- The benefits of organic mulch are that it can be collected from nature and provides cheaper crop production.
- The aim of mulching is to cut off the light to the weeds and suppress their growth.
- The crop grower should know that conventional plastic mulch is mainly used for crop production as it controls weeds, conserves soil moisture, increases soil temperature, improves crop yield and quality, has low cost and is readily available.
- Plastic mulch is around 1.0-1.5 mils thick, 4-6 feet wide, in rolls 1,000-4,000 feet long.
- Plastic mulch functions to warm the soil, conserve moisture and prevent nutrient leaching.
- An herbicide is necessary to keep weeds under control with clear mulch.
- Black mulch will prevent weed growth by prohibiting light transmittance to the soil and will warm the soil 3°-5° F over bare ground.
- Plastic mulch should be applied once the fields have been leveled, smoothed and fertilizer has been applied with good soil moisture.
- Normally, plastic mulch should be laid in spring season once land has been prepared, but certain spring seasons are wet which can delay normal land preparation and planting activities, so growers in such cases should lay plastic in the fall season.

☆ After harvesting, plastic mulches should be removed from the field and should be disposed properly as per local ordinances on incineration, landfills or open burning.

4.3 Advantages and Disadvantages of Different Types of Weeding Methods

Advantages of Plastic Mulching

1. It is completely impermeable to water.
2. It prevents the direct evaporation of moisture from the soil and thus limits the water losses and conserves moisture.
3. By evaporation suppression, it prevents the rise of water containing salts.
4. Mulch can facilitate fertilizer placement and reduce the loss of plant nutrients through leaching.
5. Mulches can also provide a barrier to soil pathogens.
6. Opaque mulches prevent the germination of annual weeds from receiving light.
7. Reflective mulches will repel certain insects.
8. Mulches maintain a warm temperature even during nighttime which enables seeds to germinate quickly and for young plants to rapidly establish a strong root growth system.
9. Synthetic mulches play a major role in soil solarisation process.
10. Mulches develop a microclimatic underside of the sheet, which is higher in carbon-di-oxide due to the higher level of microbial activity.
11. Under mulch, the soil structure is maintained during the cropping period
12. Early germination almost 2-3 days.
13. Better modulation in crops like Groundnut.
14. Fewer nematodes populations.
15. Water erosion is completely averted since the soil is completely covered form bearing action of raindrops.
16. When compared to organic mulches, it serves for a longer period.

4.4 Critical Stages of Weed Control

4.4.1 Pre-Emergence Weed Control

Introduction

☆ In India, main weeds in winter grain crops are Board left weeds such as Lamba quarters (*Chenopodium* sp.) and Wild safflower. (*Carthomus* Osycantha)

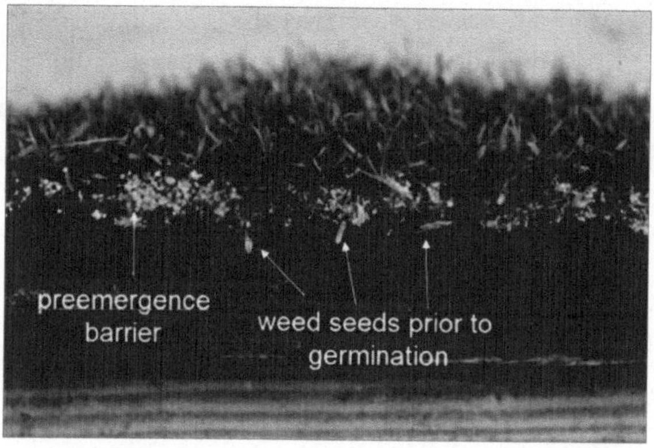

- In some parts of the country, perennial dicot weeds such as canda thistle (*Cirsium avenses*), field bindweed (*Convalvelus arvensis*) are found.
- Annual Grasses like Wild Oat (Avenu fatua), Chanry grass (Phalaris minor).
- The weed seeds get admixture in the crop seed and reduce grain quality.

Pre-Emergence Weed Killers

- These are used on established fields as part of an annual effort to combat plant pests.
- These chemical compositions are used before weeds take hold to kill off infant root systems and keep them from growing.
- It is used before seeing weeds to save them from showing up in fields that have no interference of chemicals with germination but can save the formation of new root cells in baby weed plants.
- Timing, weather and the type of problematic weeds in fields will dictate the exact formula and application for using pre-emergence procedures.
- Chemicals used in pre-emergence are not effective on vegetative buds that sprout from existing roots or rhizomes.
- The chemicals cannot be used on prepared grass seedbed as their root stunting action in young plants also affects sprouting grass.
- The grown crop has no fear, as their root system is already developed and the crop is hearty and healthy.
- The pre-emergent procedure shows that it is the sensitive root tissue of newly germinated seedlings that are killed off, which result in complete plant death.
- Perennial weeds develop thick persistent adult roots that re-sprout in spring, which makes them difficult to control with pre-emergent formula.
- The timing of the pre-emergence weed killer should match with the germination period for the variety of weed.

- Biennial weeds are not controlled by the pre-emergence procedures as they produce seed that nearly germinates the year around.
- These weed killers need water to activate themselves and carry the chemical down to the root systems of newly sprouted weeds.

4.4.2 Post-Emergence Weed Control

Introduction

- Post-emergence, attack the weeds once they grow up.
- Post-emergence herbicide controls the existing weeds and helps in preventing future ones.
- There are different types of post-emergent weed killers, what cultivator should keep information related to a variety of weeds and need to control them.
- Post-emergent weed killers attack the foliage or flow systematically to the roots of the weed.
- Herbicide application after the emergence of the crop or weed is referred to as a post-emergence application.
- When the weeds grow before the crop plants have emerged through the soil and are killed with a herbicide then it is called as early post-emergence.
- Spraying 2,4-D Na salt to control parasitic weed Strega is called post-emergence while spraying of paraquat to control emerged weeds after 10-15 days after plantation can be called as early post-emergence.
- *e.g.* Glyphosate, Paraquat, 2,4-D Na Salt.

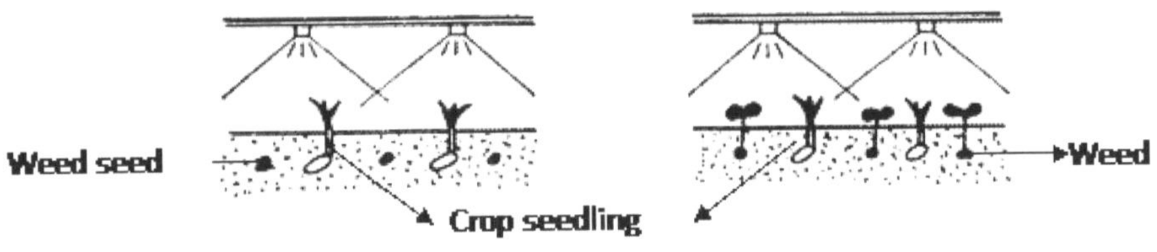

Types of Post-Emergent Weed Killers

- The post-emergent can be either systemic or contact applications.
- Systemics are useful on perennial weeds as they are absorbed directly in plant and move throughout it for maximum killing action.
- Contact herbicides kill the exposed part of plant and are used on annuals and smaller weeds which kill the entire plant.

Post Emergence Herbicides

- These herbicides are applied directly to emerged weeds.

- In contrast to pre-emergence herbicides, post-emergence herbicides control only weeds that are emerged and actively growing at the time of treatment.
- They are used for farm weed control which is systematic or translocated in the plant system that transports food and water.
- It is applied after the weeds and crops have emerged from the soil.
- The treatments can be applied in either a broadcast or directed fashion.
- When applying post-emergence herbicides, it is necessary to have maximum coverage of the weed with the spray solution.
- Surfactants are used with postemergence herbicides to enhance control.
- These herbicides are classified as selective and non-selective.
- Selective herbicides are used to target certain weeds and in areas such as turf where contact with the grass is unavoidable.
- Non-selective herbicides are used for broad weed control and have a purpose in open, unmanaged fields, for instance.
- They control weeds that have emerged since crop or pasture establishment, can be applied with little damage to the crop or pasture plants.

Benefits

- It gives a high level of target weed control with the extra benefit of improved crop or pasture yield.
- It allows fine-tuning of herbicide selection to match weeds present in the paddock.
- In this, the timing of application is flexible to suit weed size, crop growth stage and environmental conditions.
- It has pre-emergent activity on subsequent weed germinations.

Issues

- It requires careful consideration when selecting the best post-emergent herbicide in any one situation.
- The use of post-emergent herbicides to stressed crops and weeds results in reduced levels of weed control and increased crop damage.
- In this, crop competition is important for effective weed control by selective post-emergent herbicides.
- It requires a suitable technique for the situation to optimize control.
- It requires the correct adjuvant to ensure effective weed control.
- It needs a suitable formulation of herbicide for every situation.
- The effectiveness of post-emergent herbicides is influenced by a range of plant and environmental factors.

Chapter 5

Irrigation Management under Organic Farming

5.1 Timing and Method of Irrigation Appropriate for given Soil Type and Climatic Conditions

5.1.1 Irrigation Scheduling

- Irrigation scheduling is one of the factors that influences the agronomic and economic viability of small farms.
- It is important for both water savings and improved crop yields.
- The irrigation water is applied to the cultivation according to predetermined schedules based upon the monitoring of soil water status and crop water requirements
- The type of soil and climatic conditions have a significant effect on the main practical aspects of irrigation, which are the determination of how much water should be applied and when it should be applied to a given crop.
- In addition to the basic factors relevant to preparation of irrigation schedules examined, other important elements should also be considered like crop tolerance and sensitivity to water deficit at various growth stages and optimum water use.

There are many factors which crop grower needs to consider when developing an irrigation schedule for organic crops which includes:

- Water requirements of plants in each zone

- ☆ Local climate and weather
- ☆ Soil composition
- ☆ Sun exposure
- ☆ Slope
- ☆ Depth of root zone

Crop Water Requirement

- ☆ The crop type and size impact how much water are required to remain healthy in the landscape.
- ☆ Information regarding the crop water requirements for common landscape crops is provided by local utilities or extension offices or online.
- ☆ Immediately following the crop installation, landscapes at their early stages require frequent irrigation intervals to establish the plants.
- ☆ Once the landscape is established, the schedule should be adjusted to reduce the irrigation frequency appropriate for those crops/plants.

Weather Conditions

- ☆ Local climate and variations in evapotranspiration play a significant role in determining how irrigation schedules should shift over the season.
- ☆ Specific weather events like rainfall should be considered and if scheduling is done manually, hen crop grower should take such parameters into consideration and create monthly schedules.
- ☆ A variety of irrigation controllers and sensors on the market uses local conditions to determine watering times and amounts.
- ☆ Water Sense labeled weather-based irrigation controllers have been independently certified for efficiency and performance as they use local weather and landscape conditions to tailor watering schedules to actual conditions on the site.
- ☆ Soil moisture-based control technologies use moisture sensed in the soil to adapt scheduling to plant water needs.
- ☆ Sensors like rain, wind and freeze sensors, help in delay or in preventing irrigation if conditions are not ideal or necessary for watering.

Soil Types

- ☆ Soil types should be considered when developing an irrigation schedule.
- ☆ The infiltration rate of water depends on the composition of the soil.
- ☆ Coarse soil requires frequent irrigation intervals; medium loamy soil requires longer, less frequent watering while fine clay soils have a high water-holding capacity that requires the least frequent irrigation.

- ☆ If soil type is untaken into account when developing a schedule, water could be wasted due to runoff or deep percolation.
- ☆ Irrigation water on clay soil does not infiltrate quickly, and if the system is run too long, the landscape will begin to flood which results in runoff.
- ☆ To make irrigation in different soil types, including cycle and soak in irrigation schedule instead of irrigating each zone fully by using cycle and soak method irrigates in short intervals.
- ☆ If the flow rate is slow enough, there will be no pooling on the surface during the entire cycle and hence there will be no need to adjust the schedule.

Slope and Sunlight

- ☆ The slope of the landscape and its orientation to the sun should be considered while developing an irrigation schedule.
- ☆ Bulb crop growers should take care that they should not belong to irrigation schedule events on slopes, as it could result in water runoff the landscape.
- ☆ Ponding results at the lower end of landscape due to runoff from elevated areas, so lower-flow-rate emitters should be installed, and/or more frequent, shorter events should be scheduled.
- ☆ Regarding sun exposure, crop/plants exposed to intense sunlight, or on slopes with southern or western exposure, may need more irrigation due to increased evapotranspiration of the crop/plants.

Depth

- ☆ The depth of the root zone should be considered while scheduling the irrigation.
- ☆ If irrigation does not reach the full depth of the root system, crops are not receiving enough water to maximize growth.
- ☆ Both plant type and age should be considered when determining root depth and associated schedule parameters.

5.1.2 Irrigation as per Climatic Condition

- ☆ Climate change not only affects water availability but also the demand for water.
- ☆ If the climate in a given region gets drier and warmer, water availability will decrease and be exacerbated by increasing water demand.
- ☆ Soil water content and therewith irrigation in agriculture is likely to be affected by higher temperature.
- ☆ Climatic parameters changes will affect the demand for water and supply.

- ✰ Changes in water demand, irrigation practice enhance groundwater exploitation which shows that climate change impacts groundwater resources.
- ✰ As temperature increases, the atmospheric air gets warmer which accelerates the evaporation from soil moisture.
- ✰ Irrigation requirements are determined by climate and by crop and soil characteristics and are applied when the soil water content is lower than 20 per cent of available water capacity.
- ✰ There is no need for irrigation in high-rainfall as precipitation water is sufficient to grow the crop.

Climatic Factor	*Crop Water Need*	
	High	*Low*
Sunshine	sunny (no clouds)	cloudy (no sun)
Temperature	hot	cool
Humidity	low (dry)	high (humid)
Windspeed	windy	little wind

- ✰ The highest crop water needs are found in the areas which are hot, dry, windy and sunny.
- ✰ The lowest values are found when it is cool, humid and cloudy with little or no wind.
- ✰ Single crop grown in different climatic zones will have different water needs.
- ✰ For calculating water needs of the crop, it is useful to take a certain standard crops or reference crops and determine how much water is required by the crops per day in various climatic regions.

5.1.3 Appropriate Irrigation System Based on Soil Type

- ✰ It is seen that water is absorbed and it moves slowly through clay soils, but once the soil gets wet, they retain significant amount of moisture.
- ✰ Water is absorbed and moves quickly through sandy soils, but they retains very little which shows that water is applied quickly to clay soil has the tendency to run off rather than move in the soil.
- ✰ Irrigating of clay soils shows that water needs to be applied slowly over long period as a result of which, the site will not need irrigation for several days.
- ✰ Irrigation on sandy soils is applied quickly but for short periods.

- ☆ Irrigation times on sandy sites should be shorter, otherwise water moves beyond the root zone, and becomes unavailable to the plant and contributing to soil leaching.
- ☆ For efficient water use under certain weather conditions, sandy sites may need daily irrigation for short periods.
- ☆ Clay soils have greater capillary movement as compared to sandy soils.
- ☆ Quick water application on sandy soils will contribute to big wetting area which gives more soil volume for roots to exploit.

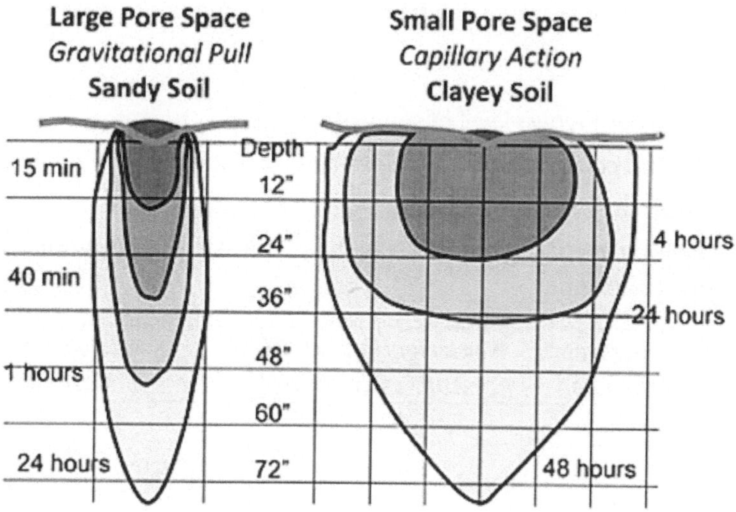

- ☆ Clay-based soils have small, flat, compact particles with large surface to volume ratios.
- ☆ Clay soil are often difficult to prepare for planting since they are slippery when wet and hard when dry, making timing for field operations critical to avoid damaging soil structure and getting proper soil tilth for planting.
- ☆ Sand-based soils are at the other end of the spectrum having comparatively large particles with small surface to volume ratios.
- ☆ Sand-based soil are easier to prepare for planting and can be worked shortly after significant rainfall.

5.1.4 Irrigation Technology as per Soil

Basin Irrigation

- ☆ It is the simplest of the surface irrigation methods.
- ☆ If the basins are small, they can be constructed by hand or animal traction.
- ☆ Their operation and maintenance is simple.

Furrow Irrigation

- It requires accurate field grading which is mostly done by machines.
- The maintenance such as ploughing and furrowing are done by machines.
- It needs skill, organization and frequent use of foreign currency for fuel, equipment and spare parts.

Short Level Furrows

- It is called as furrow basins.
- It is like basins which can be constructed and maintained by hand.

Borders

- It requires highest level of sophistication.
- They are constructed and maintained by machines.
- The grading needs to be accurate.
- Machine operation requires a high level of skill, organization and usually foreign currency.

Soil Type	*Rooting Depth of the Crop*	*Net Irrigation Depth per Application (mm)*	*Irrigation Method*
Sand	Shallow	20-30	Short furrows
	Medium	30-40	Medium furrows, short borders
	Deep	40-50	Long furrows, medium borders, small basins
Loam	Shallow	30-40	Medium furrows, short borders
	Medium	40-50	Long furrows, medium borders, small basins
	Deep	50-60	Long borders, medium basins
Clay	Shallow	40-50	Long furrows, medium borders, small basins
	Medium	50-60	Long borders, medium basins
	Deep	60-70	Large basins

5.1.5 Irrigation Scheduling Methods

- Irrigation requires a relatively high investment in terms of equipment, fuel, maintenance, and labor, but it offers significant potential for increasing net income.

- ☆ The frequency and timing of water applications have the main impact on yields and operating costs.
- ☆ To schedule irrigation for the most efficient use of water and to optimize production, crop growers should frequently determine the soil water conditions throughout the root zone of the crop being grown.
- ☆ Many methods have developed and are used that compares investment in irrigation equipment.
- ☆ When properly used and coupled with grower experience, a scheduling method improves the irrigator's chances of success.

1. Water Balance Method

- ☆ In the water balance method, the idea is to get balance of incoming and outgoing soil water so that adequate available water is maintained for the plant.
- ☆ In this, input includes incoming water in any form whether rainfall or irrigation while output includes any type of water removal which is known as evapotranspiration.
- ☆ Evapotranspiration consists of water removal by plant and water loss due to evaporation which uses two variations of water balance method.
- ☆ To use a variation, crop growers must know the soil type and available water holding capacity of soil obtained for soil from the soil conservation guide.
- ☆ The grower should determine the zone which is trying to manage where the zone varies as per the effective rooting depth of a particular crop.
- ☆ In this, effective rooting depth of certain crops may be less, so determining the total water, grower has available to manage in such a zone.
- ☆ It is desirable to try to manage the only percentage of total water.
- ☆ As water is removed daily, such amounts are subtracted from the adjusted water available column.
- ☆ When water available approaches zero balance, the grower should irrigate where the amount added depends on soil type which will be the same as 50 per cent.

2. Soil Moisture Measuring

- ☆ Many devices like tensiometer, electrical resistance meters are available for monitoring or measuring soil moisture.
- ☆ The most common in use in the field are tensiometers, moisture blocks and various electronic soil probes.

- The expense of device results as a factor, so it is important for crop growers to make sure of all necessary inputs.

3. Computer Models

- Computer models are developed and updated each year.
- To use a computer model, the grower should have access to a computer on a regular basis.
- Simple models are available that do checkbook type record keeping.
- Many sophisticated models are present that needs weather data inputs.
- Models also incorporate other management decisions like whether to spray for insects and diseases.
- Models of later type are known as expert systems.
- Computer models are beneficial in decision making, but Garbage in – Garbage out is applicable in computer models.

4. Keeping a Chart

- Crop growers can get the benefit from using tensiometers, electrical meters or computer models by recording soil moisture readings and plotting them on a chart.
- The grower can plot the reading directly in the field.
- Growers should use different colored pencils for different depth tensiometers for making the chart easier to read where lines show activities in the past.

5.2 Quantity of Water Required for a Specific Crop and its Effect on its Yield

5.2.1 Need of Water for Crop

- Water is one of the most important inputs essential for the production of crops.
- Plants need water continuously during their life in huge quantities.
- It influences photosynthesis, respiration, absorption, translocation, and utilization of mineral nutrients, and cell division apart from other processes.
- Water shortage and excess will affect the growth and development of a plant directly and, consequently, its yield and quality.
- Water affects the performance of crops not only directly but also indirectly by influencing the availability of other nutrients, the timing of cultural operations, *etc.*

- ☆ Water and other production inputs interact with one another.
- ☆ In proper combinations, crop yields can be boosted manifold under irrigated agriculture.
- ☆ It is a costly input when canals supply it.
- ☆ The constructing of dams and reservoirs, the conveying of water from storage points to the fields, the operating and the maintaining of canal systems involve huge expense.
- ☆ The misuse of water leads to problems of waterlogging, salt imbalance, *etc.* which render agricultural lands unproductive.
- ☆ A proper appreciation of the relationship and economic utilization of water resources is required for maximum crop production.

5.2.2 Optimum Water Level

- ☆ Different crops need a different type of soils, amounts of nutrients and different types and amounts of water.
- ☆ The amount of water required by the plant is dependent on the growing season and the climate where it is grown.
- ☆ Selecting the right crop for given soil conditions and climate will save water requirements for irrigation.
- ☆ The water need of a crop is expressed in mm/day, mm/month or mm/season.
- ☆ If the water need of certain crops in a very hot, dry climate is 10 mm/day, then each day the crop needs a water layer of 10 mm over the whole area on which the crop is grown.
- ☆ The crop water need is the depth of water required to meet the water loss by evapotranspiration.
- ☆ The crop water need refers to crop grown under optimal conditions, *i.e.* a uniform crop, actively growing, completely shading the ground, free of diseases, and having favorable soil conditions.

5.2.3 Water Requirement for Crops

- ☆ It is essential to know the water requirement of a crop which is the total quantity of water required from its sowing time up to harvest.
- ☆ Naturally different crops may have different water requirements at different places of the same country, depending upon the climate, type of soil, method of cultivation, effective rain, *etc.*
- ☆ The total water required for crop growth is not uniformly distributed over its entire life span known as crop period.

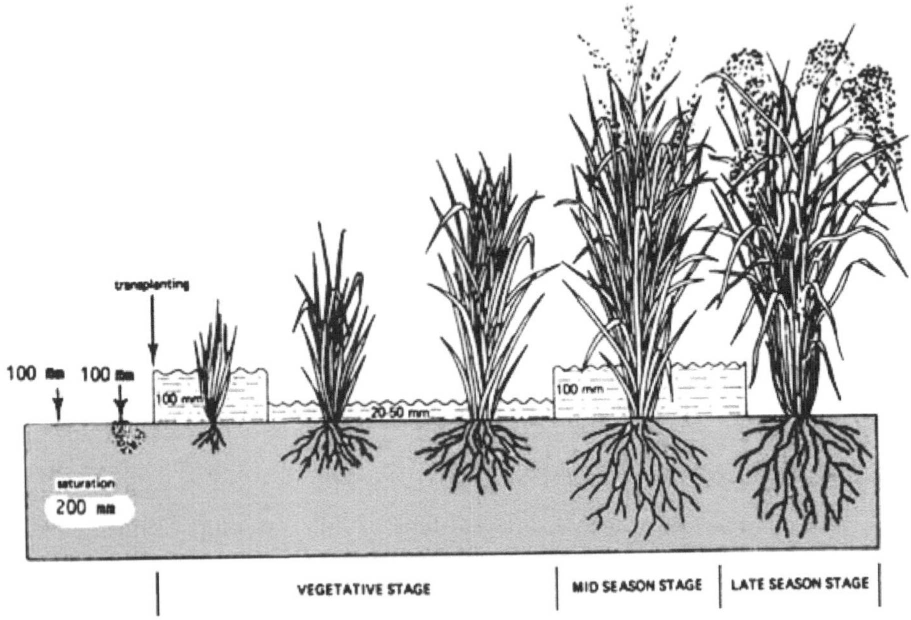

- ☆ The watering stops same time before harvesting and time duration from first irrigation at the time of sowing to last before harvest is known as base period.
- ☆ As water is essential for crop growth, its requirement for each crop is definite which differs as per the season or as per soil type but net requirement results same.

5.2.4 Total Water Requirement in Growing Crop

The water that is required to irrigate a field or plot of land growing the particular crop not only has to satisfy the evapo-transpiration needs for growing the crop, but also include:

- ☆ Losses in form of deep percolation while conveying water from inlet of the field till its last or tail end as water gets distributed within the field
- ☆ Water requirement for special operations like land preparation, transplanting, leaching of salts, *etc.*
- ☆ Evapo-transpiration requirement of crops will not include water required by crops for building up plant tissues, as compared to evaporation needs. *etc.* is often related as the consumptive irrigation requirement.

The net irrigation requirement is the amount of irrigation water required to be delivered in the field to meet the consumptive requirement of crop and other needs like:

- ☆ Leaching,
- ☆ Pre-sowing
- ☆ Nursery water requirement

Thus, NIR (net irrigation requirement) = CIR + LR + PSR + NWR

where,

- ☆ LR = Leaching requirement
- ☆ PSR = Pre-sowing requirement
- ☆ NWR = Nursery water requirement

The total depth of water required to raise a crop over a unit area of land is known as delta which has some typical values for common crops such as:

Rice

- ☆ 1000mm to 1500mm for heavy soils or high water table
- ☆ 1500mm to 2000mm for medium soils
- ☆ 2000 to 2500 for light soils or deep water table
- ☆ 1600mm for upland conditions

Wheat

- ☆ 250mm to 400mm in northern India
- ☆ 500mm to 600mm in Central India
- ☆ Barley: 450mm

Maize

- ☆ 100mm during rainy season
- ☆ 500mm during winter season
- ☆ 900mm during summer season
- ☆ Cotton: 400 – 500mm

Sugarcane

- ☆ 1400mm to 1500mm in Bihar
- ☆ 1600mm to 1700mm in Andhra Pradesh
- ☆ 1700mm to 1800mm in Punjab
- ☆ 2200mm to 2400mm in Madhya Pradesh
- ☆ 2800mm to 3000mm in Maharashtra

5.2.5 Irrigation Requirement of Crops

Crop	*Growing Period (days)*	*Total Water Requirement*		*Daily Water Requirement*	
		in cm	*in inches*	*in cm*	*in inches*
Jawar	114	64.25	25.70	0.575	0.23
Maize	100	44.50	17.80	0.450	0.18
Rice	93	104.50	41.80	1.075	0.43
Wheat	88	37.00	14.80	0.425	0.17
Groundnut	124	65.25	26.10	0.525	0.21
Linseed	88	31.71	12.68	0.350	0.14
Cotton	202	105.50	42.20	0.525	0.21
Sugarcane	365	237.50	95.00	0.650	0.26
Tobacco	132	98.00	39.20	0.750	0.30
Onion	120	75.00	30.00	0.625	0.25
Potato	88	30.00	12.00	0.750	0.30
Pea	88	30.00	12.00	0.350	0.14
Mustard	88	25.20	10.08	0.300	0.12
Barley	88	25.20	10.08	0.400	0.16
Oat	88	36.00	14.40	0.400	0.16
Ragi	127	74.50	29.80	0.575	0.23

5.2.6 Ample/Shortage of Water Availability

- If the water available is adequate then irrigation should be given at all the stages.
- If water is in short supply then irrigations should be given at certain critical stages.
- Maximum reduction takes place if the irrigation at the critical growth stage is missed.
- If the water is available for less than adequate number (5-6) of irrigations then crop should be irrigated as:
  - For one irrigation: it should be applied at crown root initiation in 20-25 days after sowing.
  - For two irrigations: first irrigation at crown root initiation and second at flowering stage.
  - For three irrigations: first irrigation at crown root initiation, second at late jointing and third at milk stage.

- For four irrigations: crop to be irrigated at crown root initiation, late jointing, heading and milk stage.

5.3 Frequency of Irrigation Required at Various Stage of Plant Growth

5.3.1 Growth Stage of Crop as per Water Required

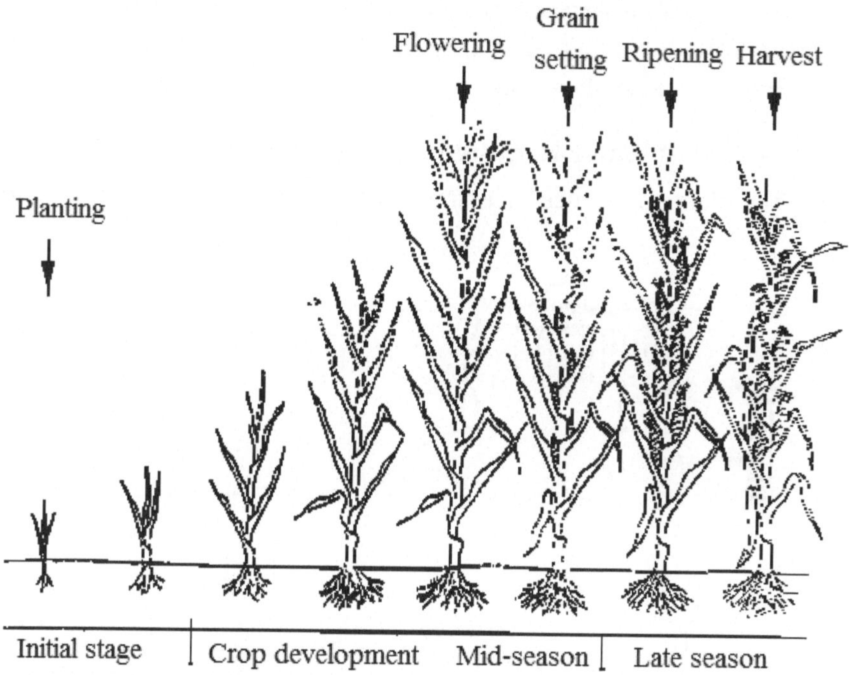

- A fully grown crop will need more water than the crop which has just been planted.
- The crop water need or crop evapotranspiration consists of transpiration by the plant and evaporation from the soil and plant surface.
- If crop are very small, the evaporation will be more important than the transpiration.
- If the plants are fully grown the transpiration is more important than the evaporation.
- At planting and during the initial stage, evaporation is important than transpiration and the evapotranspiration or crop water need during initial stage is estimated at 50 per cent of crop water required during the mid - season stage, when the crop is fully developed.
- At crop development stage, crop water requirement gradually increases from 50 per cent of maximum crop water to maximum crop water need.

- ☆ The maximum crop water need is reached at the end of the crop development stage which is the beginning of the mid-season stage.
- ☆ At late season stage, the period during which the crop ripens and is harvested, a distinction can be made between fresh harvested crops and dry harvested crops
- ☆ With fresh harvested crops, water requirement remains the same during the late season stage as it was during the mid-season stage.
- ☆ With dry harvested crops at late season stage, these crops are dried out and even die, so their water needs during the late season stage is minimal.

5.3.2 Quantity of Water Required According to Stages of Growth

- ☆ The plant roots suck or extract water from the soil to live and grow.

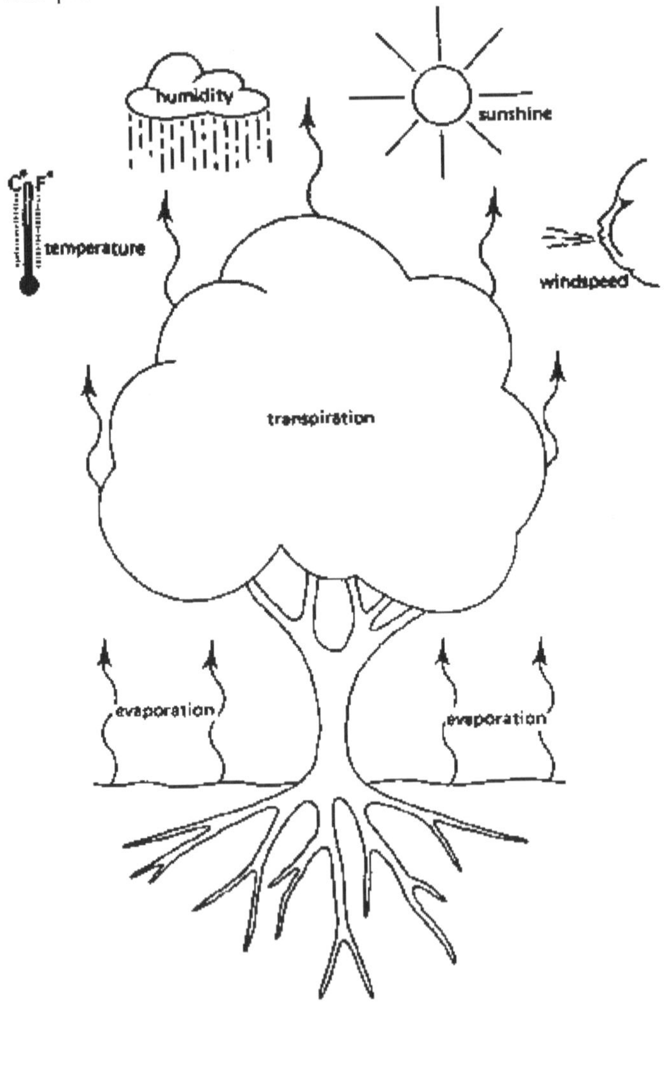

- The main part of the water does not remain in the plant, but escapes to the atmosphere as vapor through the plant's leaves and stem which is known as transpiration.
- Transpiration happens mainly during the day time, where water from an open water surface escapes as vapor to the atmosphere during the day which same happens to water on the soil surface and to water on leaves and stem of a plant.
- The water need of a crop comprises of transpiration and evaporation known as evapotranspiration.
- The water need of a crop is usually expressed in mm/day, mm/month or mm/season.
- A certain crop grown in a sunny and hot climate needs per day more water than the same crop grown in a cloudy and cooler climate.
- Apart from sunshine and temperature, other climatic factors which influence the crop water need includes humidity and wind speed.
- When it is dry, the crop water needs are higher than when it is humid.
- In windy climates the crops will use more water than in calm climates.

5.3.3 Irrigation Schedule under Limited Supply of Water

With limited supply of water, following irrigation schedule of irrigations should be adopted which strongly stresses importance of irrigation at CRI stage:

- Where only one irrigation is possible, give crown root initiation CRI stage 20-25 days after sowing.
- Where two irrigation are available, first irrigation be given at CRI stage and second flowering stage.
- Where three irrigation are possible, first irrigation be given at CRI stage while second irrigation at late jointing stage and third irrigation at milk stage.
- It shows that each week delay in first irrigation from CRI stage results in yield reduction of 2-3 quintal per hectare.

5.4 Different Types of Irrigation System Available

5.4.1 Types of Irrigation System

The irrigation system consists of:

Intake Structure

- The intake structure is built at the entry to the irrigation system.
- The purpose of intake structure is to direct water from the original source of supply (lake, river, reservoir *etc.*) into the irrigation system.

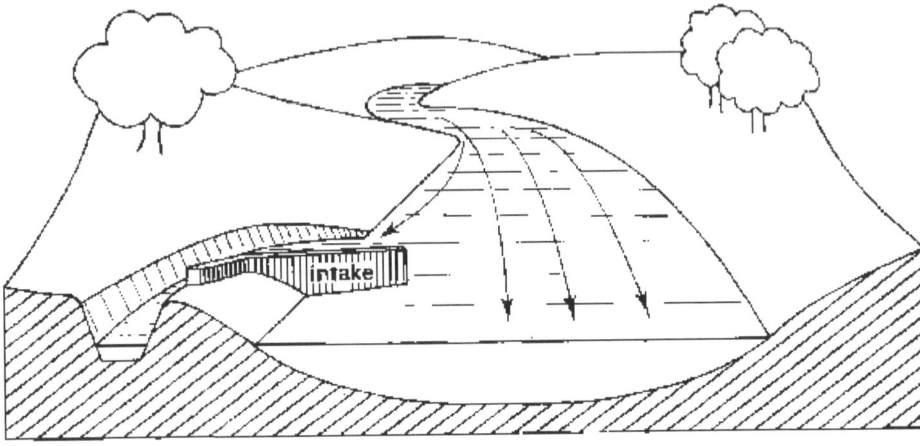

Pumping Station

- Many times, irrigation water source lies below the level of the irrigated fields.
- The pump is used to supply water to the irrigation system.
- There are several types of pumps, but the most commonly used in irrigation is the centrifugal pump.

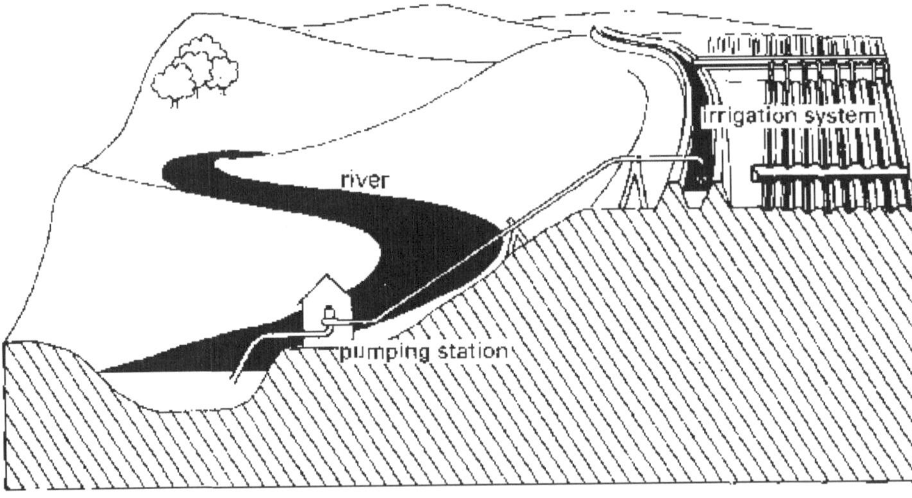

Conveyance System and Distribution System

- The conveyance and distribution systems consist of canals transporting the water through the whole irrigation system.
- Canal structures are required for the control and measurement of the water flow.

Open Canals

- ✰ An open canal, channel, or ditch, is an open waterway whose purpose is to carry water from one place to another.
- ✰ Channels and canals refer to main waterways supplying water to one or more farms.
- ✰ Field ditches have smaller dimensions and convey water from the farm entrance to the irrigated fields.
- ✰ The most commonly used canal cross-section in irrigation and drainage, is the trapezoidal cross-section.

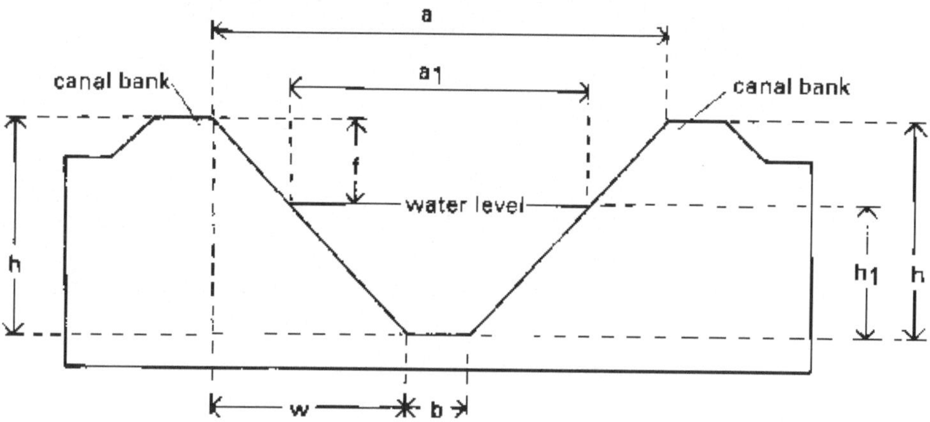

a = top width of the canal
a_1 = top width of the water level
h = height of the canal
h_1 = heigth or depth of the water in the canal
b = bottom width of the canal
h:w = side slope of the canal
f = free board (= $h-h_1$)

Field Application System

- ✰ The traditional methods of applying water to the field is by bringing the water from the source of supply, such as a well, to each plant with a bucket or a water-can.
- ✰ Surface irrigation is application of water to the fields at ground level in which, either the entire field is flooded or the water is directed into furrows or borders.
- ✰ Furrows are narrow ditches dug on the field between the rows of crops.

- ☆ The water runs along the furrow as it moves down the slope of the field.
- ☆ The water flows from the field ditch into the furrows by opening up the bank or dyke of the ditch or by means of siphons or spiels.

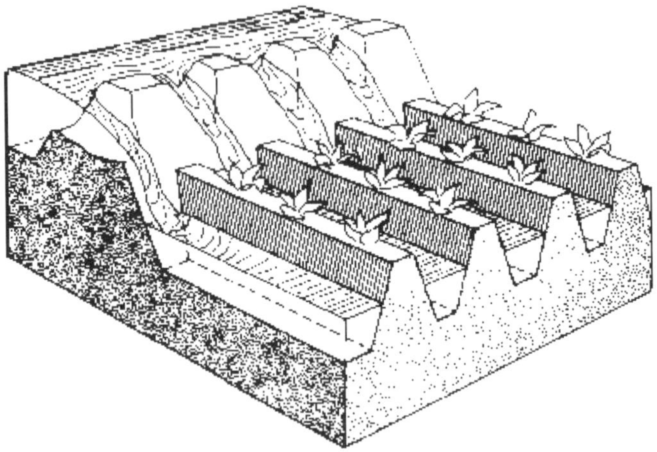

- ☆ Siphons are small curved pipes that deliver water over the ditch bank.

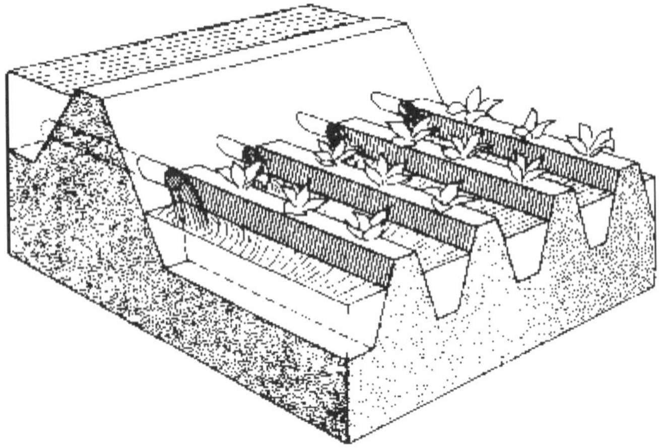

- ☆ Spiels are small pipes buried in the ditch bank.
- ☆ In border irrigation, the field to be irrigated is divided into strips by parallel dykes or border ridges.

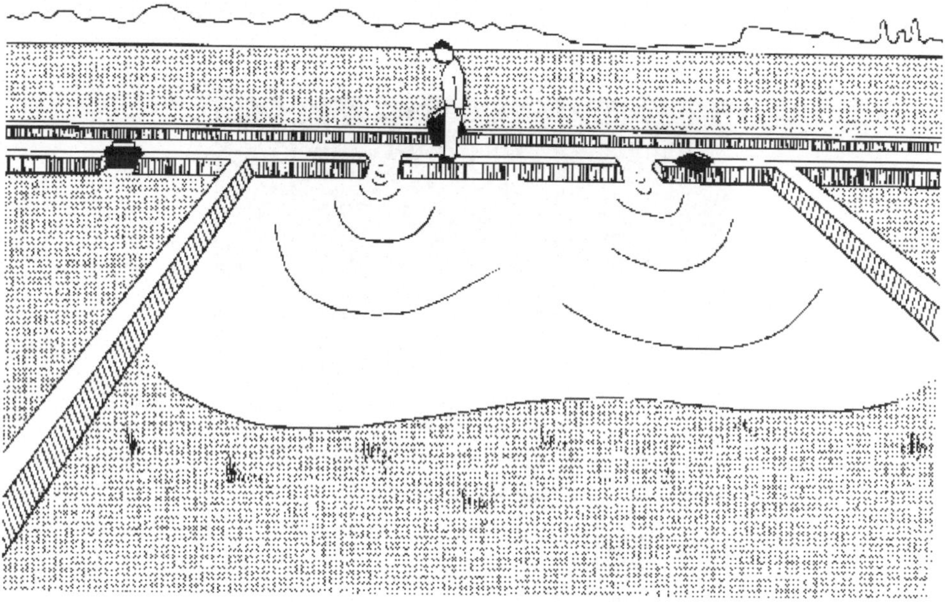

- ☆ In border irrigation, water is released from the field ditch on border through gate structures called outlets or by means of siphons or spiels.
- ☆ In basin irrigation, basins are horizontal, flat plots of land, surrounded by small dykes or bunds in which the banks prevent the water from flowing to the surrounding fields.

☆ Basin irrigation is commonly used for rice grown on flat lands or in terraces on hillsides.

☆ Trees can also be grown in basins, where one tree usually is located in the centre of a small basin.

Drainage System

☆ A drainage system is required to remove excess water from the irrigated land.

☆ The excess water may be *e.g.* waste water from irrigation or surface runoff from rainfall.

☆ Waste water can also be leakage or seepage water from the distribution system.

☆ Excess surface water is removed through shallow open drains.

☆ Excess groundwater is removed through deep open drains or underground pipes

5.4.2 Micro Irrigation System

Introduction

☆ It is the slow application of water as discrete or continuous drips, tiny streams or miniature spray system on, above, or below the soil by surface drip.

☆ It is applied through emitters connected to water delivery line by low-pressure delivery.

- ☆ It has very small openings known as emitters that are designed to discharge the water at a known rate for operating pressure.
- ☆ It needs more planning than overhead sprinkler systems as the emitter systems deliver water directly to the root system.
- ☆ It encompass several ways of water application to the crop such as:
  - ❖ Drip irrigation
  - ❖ Spray irrigation
  - ❖ Subsurface irrigation
  - ❖ Bubbler irrigation

Types of Micro Irrigation Systems

1. Drip Irrigation

Components

- ☆ Pump station
- ☆ By-pass assembly
- ☆ Control valves
- ☆ Filtration system
- ☆ Fertilizer tank/Venturi
- ☆ Pressure gauge
- ☆ Mains/Sub-mains
- ☆ Laterals
- ☆ Emitting devices
- ☆ Micro tubes

- ☆ It is also known as trickle irrigation system.

- ☆ It is frequent, slow application of water to soils by using mechanical devices known as emitters or applicators which are located at selected points along the delivery lines.
- ☆ The emitters dissipate the pressure from the distribution system using orifices, vortexes and tortuous or long flow paths which allow limited volume of water to discharge.
- ☆ The emitters are placed on the ground or are buried.
- ☆ In this, emitted water moves in the soil system mainly by unsaturated flow.
- ☆ The wetted soil area for widely spaced emitters is normally elliptical in shape.
- ☆ In this, the water is led to the field by pipe system where next to row of plants or trees, a tube is installed.
- ☆ At regular intervals, near the plants or trees, a hole is made in the tube and equipped with an emitter.
- ☆ The water is supplied slowly, drop by drop, to the plants through these emitters

Advantages

- ☆ Increases production and productivity.
- ☆ Improves quality and ensure early maturity of crops.
- ☆ Water Saving up to 40 per cent - 70 per cent.
- ☆ Controls weed growth, saving of fertilizer (30 per cent) and labour cost (10 per cent).
- ☆ Fertigation/Chemigation can be made efficiently.
- ☆ Control diseases.
- ☆ Use of saline water is possible.
- ☆ Soil erosion is eliminated.
- ☆ Suitable for uneven/undulating land.
- ☆ High Water Use Efficiency

2. Sprinkler irrigation

- ☆ It creates artificial rainfall, where the water is led to the field by pipe system where water is under pressure.
- ☆ In this, the spraying is done by using many rotating sprinkler heads or spray nozzles or single gun type sprinkler.
- ☆ In this, the sprinklers deliver the water to the plant and/or soil which takes water from the source and distribute the water uniformly over an area in droplet form.

- To cover larger area, sprinkler needs to throw water to a considerable distance.
- If the sprinkler is good, it will take many factors into account which includes water supply, soil, crop, topography, and atmospheric conditions.

Spray-type Sprinklers

- The spray nozzles require considerably less pressure and so energy than impact sprinklers.
- Spray nozzles are installed at the end of drop tubes/hoses to release water closer to the crop canopy to reduce wind and evaporative losses.
- The spray nozzle has a smaller wetted diameter that causes very high application rates at the outer end of the lateral.
- In heavy soil, it leads to application rates that exceed the soil infiltration rate which results in runoff.

5.5 Relative Advantages and Disadvantages of Irrigation Equipments

Watering Can

- It provides a simple and accessible irrigation technique that is understandable and widely practiced by small-scale farmers for crop production.
- The technique needs low investments, but is labor-intensive and allows irrigation of the only small area.

Treadle Pumps

- It is a micro-irrigation technique that needs relatively modest investment.
- It allows the small-scale farmers to irrigate more substantial areas than is possible with traditional watering can method.
- It is of two main types; pressure treadle pump and gravity treadle pump.
- The treadle pump is successful in wetland development where groundwater can be found at shallow depth.
- The pressure treadle pump can be connected by a set of the low-pressure pipe distribution system and flexible hoses to distribute water by sprinkling.

Motorized Pumps

- It can access to a regular supply of fuel at an affordable price.
- It needs high investment costs.
- Their result of management problems in large pump schemes
- It leads to low irrigation efficiencies due to unfamiliarity with water conveyance and field irrigation practices.

Solar Pumps

- Panels and suitable pumps are commercially available.
- Construction of a reservoir for 2 to 3 days storages to increase discharge and periods of low sunshine
- Low-pressure pipe system or drip irrigation
- Competent technical advisory services for design and installation
- It needs high investment costs.
- With this, small field areas can be irrigated.

Pump

- Pumps for irrigation simplify design and construction and eliminate long intake canals of diversion schemes.

- It provides greater flexibility in scheme size and the selection of suitable locations near the area to be irrigated.
- The energy costs of running the pumps, off-set the reduced investment costs and proved major constraint in small-scale pump schemes.
- The pumps mobility and ability easily adapt to varying water depths.
- The installation costs of the pump are lower than those of diversion intake with the long head canal, but fuel and repair costs are constraints in many small-scale irrigation schemes.

Head Control Unit

- It consists of shut-off, air, and check valves to control the discharge and pressure of water in the entire system.
- A pressure relief valve is installed after the pump unit to return excess water when the system is not operated at its full capacity.
- It may also have filters to clear the water.
- The filters remove sediment and debris, which can clog the system.
- Disc filters are commonly used to filter water from ponds, reservoirs, tanks and other sources that contain algae.
- Some head control units contain a fertilizer or nutrient tank to supply fertilizer solution to plants.

Emitters or Drippers

- These devices are used to control the discharge of water from the laterals to plants.
- They are made of High-Density Polyethylene plastic.
- Water enters the drippers at approximately 1 kg/cm2 pressure and is delivered at zero pressures in the form of droplets at a low rate of 1–2.4 liters/hour.
- There are mainly two types of emitters.

Exhaust Header

- The exhaust header serves to equalize pressure and flow between runs of inline drip tubing and provide an outlet path for flushing.
- The exhaust header does not need to be sized to equal the supply header but must be able to accommodate the flow rate of the flush valve, without exceeding 5 FPS.

Flush Valves and PLD Cap

- Flush valves are required on every inline drip system and must be located to provide an outlet from every point in each zone.

- ☆ Regular flushing not only removes debris and particles from the tubing, but the high volume of turbulent water during flushing also helps dislodge biological growth.
- ☆ Flush valves can be either automatic or manual.

Linear move Sprinkler

- ☆ Linear move sprinkler irrigation system is similar to the center pivot system in construction, except that neither end of the lateral pipeline is fixed.
- ☆ It is composed of a series of towers suspended and move laterally in the direction of rows.
- ☆ The whole line moves down the field perpendicular to the lateral.
- ☆ Water delivery to the continuous moving laterally is by a flexible hose or open ditch pickup.
- ☆ Both the center pivot and linear move systems are capable of high-efficiency water applications.
- ☆ By 'water efficiency', it means reducing water wastage by measuring the amount of water required for a specific purpose and the amount of water delivered or used.
- ☆ Such a system requires high capital investment but is not labor-intensive.

Laterals

- ☆ The laterals deliver water from mainlines/sub-main lines to the sprinklers.
- ☆ They can be permanent but more often they are portable and made of aluminium alloy or plastic so that they can be moved easily.
- ☆ In a portable system, laterals need to be of the same size so that they can be changed easily.

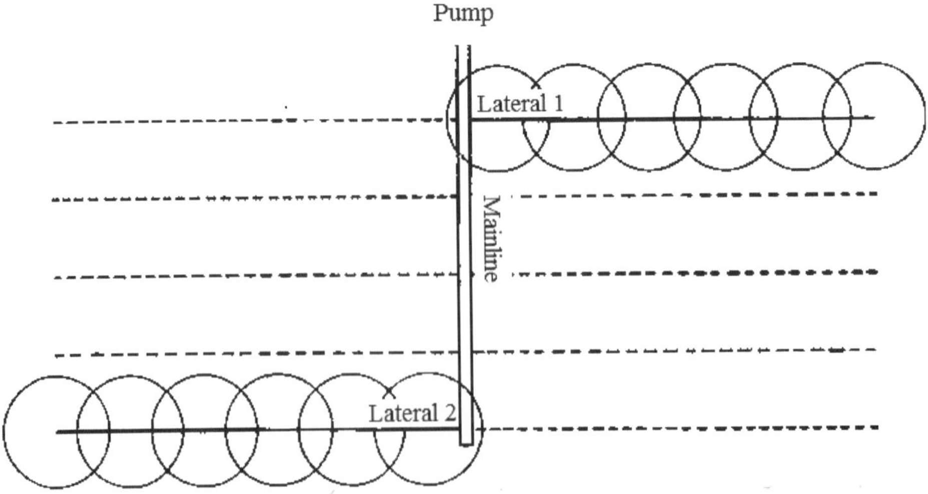

- In this, lightweight aluminium or plastic pipes are moved by hand.
- In this, rotary sprinklers are spaced 9-24 m apart along the lateral which is normally 5-12.5 cm in diameter.
- The lateral pipe is located in the field until the irrigation is complete.
- The pump is switched off and lateral is disconnected from the mainline and moved to the next location where it gets re-assembled and connected to mainline which starts irrigation again.
- The lateral can be moved 1-4 times in a day and can be gradually moved around the field till the whole field is irrigated.

Sprinklers

- The type of sprinkler depends on its nozzle size and pressure with which it discharges water.
- It must also be ensured that the water discharged does not cause run-off or damage to the crops.
- Sprinkler should supply water to the crops sown in a field uniformly under the prevailing wind conditions.
- It must meet the irrigation water requirement of a crop and irrigation frequency.
- The common symbols used when designing sprinkler irrigation plans can be:
  - Sprinkler – full: It is a nozzle which throws water all around it at 360° and at a distance of 3.6–4.5 metre.
  - Sprinkler – half: It is a nozzle which throws water all around it at 180° and at a distance of 3.6–4.5 metre.
  - Sprinkler – quarter: It is a nozzle which throws water all around it at 90° and at a distance of 3.6–4.5 metre.

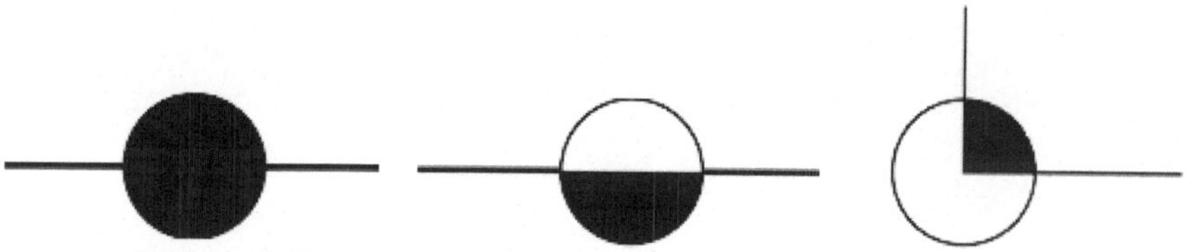

sprinkler – full *sprinkler – half* *sprinkler – quarter*

  - Sprinkler – one-third: It is a nozzle which throws water all around it at 120° and at a distance of 3.6–4.5 metre.
  - Sprinkler – three quarter: It is a nozzle which throws water all around it at 270° and at a distance of 3.6–4.5 metre.

- ❖ Sprinkler – two-third: It is a nozzle which throws water all around it at 240° and at a distance of 3.6 to 4.5 metre.

Water Meters

- ✰ It is essential for micro irrigation technician to monitor flow in order to monitor the operation of the irrigation system and crop's water use.
- ✰ The drip system is designed to produce a specific flow rate at given pressure.
- ✰ Changes in flow rate may indicate leaks in the system, improperly set pressure regulating valves or even changes in the well and pumping plant, so micro irrigation technician has to make use of water meter and pressure gauge so as to determine the drip system problems.

Pressure Gauges

- ✰ The micro irrigation technician should use pressure gauges in order to ensure that the drip system, filter system and pump operates at correct pressure.
- ✰ Pressure gauges are also critical to assess potential problems with the system.

Chapter 6

Integrated Pest and Disease Management under Organic Farming

6.1 Mixed Cropping Techniques for Pests

6.1.1 Introduction

- It is known as polyculture, inter-cropping or co-cultivation where planting of two or more plants is done simultaneously in the same field.
- In mixed cropping, if one crop faces adverse conditions or pathogen attack then another crop can be saved.
- The cropping is best suited for small-scale farming where harvesting is done by hand.
- It is successfully used to improve income and food production for small farmers and lessen the likelihood of total crop failure as if one crop fails, other crops in the field will still produce.
- The cropping requires fewer nutrient inputs like fertilizers, pruning, pest control and irrigation as compared to monoculture farming.
- The cropping is a mixture of legumes, cereals or tuber crops which is common in marginal agro-ecological environments and fulfills complimentary use of growth factors like soil nutrients, light, and water, reduces pest and disease incidence, reduced soil erosion, more total biomass production, more yield stability, and more household food security.

- For farmers with small fields, such cropping provides food security as few crops will produce some yield even if conditions are not optimum for a single crop.
- In this, the total yield per unit area is more as compared to monocropping system with ample crop cover to protect soil from erosion.
- The cropping systems offer a yield buffering capacity by diverse growing demands and different periods of root, leaf and seed development of plant varieties.

6.1.2 Criteria for Selection of Crops

- In this, crops should have different maturity times.
- Crops grown in the same field should have different water requirements.
- The nutrient requirement of one crop should be lesser than other crop.
- In this, if one crop is deep-rooted, then the other crop should have shallow roots.
- In this, if one crop is tall, then the other crop should be dwarf.

6.1.3 Advantages

- It is the practice of cultivation of two or more crops simultaneously on the same piece of land.
- It minimizes risk and insures against crop failure due to abnormal weather conditions.
- The cropping is more cost-effective.
- Increase in yield
- Optimum utilization of soil
- Minimum pest infestation
- No risk of crop failure
- Farmers can harvest varieties of produce at the same time

6.1.4 Uses

As crops ripen during different seasons, planting of more than one crop will save space and provides a wealth of environmental benefits like:

- Maintaining balance of input and outgoing of soil nutrients, weed, disease
- Insect pest suppression
- Resistance to climate extremes
- Increase in overall productivity
- Management of scarce land resources to maximum potential

6.1.5 Mixed Cropping and Pests

- Mixed cropping such as row and strip intercropping are important tools for pest and disease management in organic farming.
- The cropping has been shown to be an effective disease management tool, especially in cereals.
- In such cropping, if crops like winter-rye/winter-wheat and spring-barley/ oats are grown in the same fields, then they will reduce fungal leaf diseases.
- By using mixed cropping techniques, pests get confused as it recognizes suitable crops either by sight or by smell, but a mixture of sight and smell will serve as a structure to confuse the pest.
- Mixing cropping of species will help in pest control by influencing the availability of light which affects insect behavior and creates a barrier.

6.2 Trap Crops for Pests

6.2.1 Trap Crop

- Trap crops are grown to attract insects or other organisms to protect main crops from pest attack.
- The idea behind trap cropping is that all pests show distinct preference to certain crop stage.
- Protection can be done either by preventing pests from reaching the crop or concentrating them in certain parts of the field where they can be damaged.
- The technique is adopted by growers due to difficulties in cropping with pest situations in other ways.

Examples

- The trap crop bhindi/okra in cotton traps the bollworms in 1:10 ratio and marigold at the border of the field.
- Sesamum is used as a trap crop to attracts Diamondback moth in case of cabbage and cauliflower and to trap the pest, plantation should be of 2 rows of sesamum for every 25 rows of cabbage or cauliflower.
- In the case of groundnut, castor or sunflower can be used to attract leaf-eating caterpillar on the border of the field.
- The tomato marigold or cucumber is used as a trap crop for every 15 rows of the main crop to attract tomato fruit borer.
- Infield beans, chrysanthemum will act as a trap crop against leaf miners.
- To trap cornstalk borer in maize, sorghum has been exploited as a trap crop.
- Bihar hairy caterpillar in cowpea gets trapped by planting Gingelly.

6.2.2 Features

- The trap crop needs to be attractive to pest then the main crop.
- Trap crops should occupy a small area as far as possible and should be established early/later/along with the main crop.
- It significantly reduces pesticide usage for controlling leaf-footed bugs.
- With this, growers can reduce their inputs such as fuel, labor, time and insecticides which lead to an increase in income while protecting pollinators and other beneficial insects.

6.2.3 Trap Crops to Reduce Pests

- It is a technique that is used for pest species that move infield from edges with strong preferences for particular species or varieties of hosts that need months for reproduction.
- To trap insects, the grower should plant trap crops all around the main crop.
- When pest arrives in the field from the edges, it encounters the trap crop first and stops, staying along the edge of the field.
- The trap crop is sprayed with insecticide or flamed to kill pests before it moves to the main crop.

6.2.4 Trap Cropping Practice

Trap Crop	*Main Crop*	*Method of Planting*
Alfalfa	Cotton	Strip intercrop
Basil and marigold	Garlic	Border crops
Castor plant	Cotton	Border crop
Chervil	Vegetables Ornamentals	Among plants
Chinese cabbage, mustard and radish	Cabbage	Planted in every 15 rows of cabbage
Beans and other legumes	Corn	Row intercrop
Chick pea	Cotton	Block trap crop at20 plants/sq m
Collards	Cabbage	Border crop
Corn	Cotton	Row intercrop, planted inevery 20 rows of cottonor every 10-15 m
Cowpea	Cotton	Row intercrop in every 5 rows of cotton

Trap Crop	Main Crop	Method of Planting
Desmodium	Corn Cowpea Millet Sorghum	Row intercrop
Dill and lovage	Tomato	Row intercrop
Green beans	Soybean	Row intercrop
Horse radish	Potato	Intercrop
Hot cherry pepper	Bell pepper	Border crop
Indian mustard	Cabbage	Strip intercrop in between cabbage plots
Marigold	Solanaceous Crucifers Legumes Cucurbits	Row/strip intercrop
Medic, *Medicago litoralis*	Carrot	Strip intercrop in betweencarrot plots
Napier grass	Corn	Intercrop Border crop
Nasturtium	Cabbage	Row intercrop
Okra	Cotton	Border crop
Onion and garlic	Carrot	Border crops or barriercrops in between plots
Radish	Cabbage family	Row intercrop
Rye	Soybean	Row intercrop
Sesbania	Soybean	Row intercrop at a distanceof 15 m apart
Sickle pod	Soybean	Strip intercrop
Soybean	Corn	Row intercrop
Sudan grass	Corn	Intercrop Border crop
Sunflower	Cotton	Row intercrop in every5 rows of cotton
Tansy	Potato	Intercrop
Tobacco	Cotton	Row intercrop, planted inevery 20 rows of cotton
Tomato	Cabbage	Intercrop (Tomato planted2 weeks ahead at plots' borders)
Vertiver grass	Corn	Perimeter crop

6.3 Natural Enemies of Pests

6.3.1 Introduction

- ☆ Classical biological control introduces pest's natural enemies to locations where they do not occur naturally, but augmentation involves the supplemental release of natural enemies, boosting the naturally occurring population and conservation that involves in conservation of existing natural enemies in environment.
- ☆ Natural enemies of insect pests, known as biological control agents, are predators, parasitoids and pathogens.
- ☆ Crop pests include insects, weed, plant pathogens, invertebrate, and vertebrate animals. Identification of insect pests and their natural enemies is an important step in any pest management program.
- ☆ Insect pests and natural enemies could be identified using keys and field guides or otherwise consulting an official identification bodies.
- ☆ Unlike insect pests, plant pathogens including fungi, bacteria, virus, and nematodes are difficult to identify in the field and may need laboratory diagnosis.
- ☆ Predators are lady beetles and lacewings which are free-living species that consumes large number of prey during their lifetime.
- ☆ Parasitoids are species whose immature stage develops on or in single insect host that kills the host.
- ☆ Most parasitic insects are either flies (Order Diptera) or wasps (Order Hymenoptera).
- ☆ Pathogens are disease causing organisms like bacteria, fungi and viruses which kills or debilitate their host and are specific to certain insect groups.
- ☆ With naturally occurring disease outbreak (epizootic), there are certain beneficial pathogens which are commercially available as biological or microbial pesticides such as Bacillus thuringiensis or Bt, entomopathogenic nematodes and granulosis viruses.
- ☆ Life cycles of natural enemies can be simple or complex from which not all natural enemies of insects are beneficial to crop production.
- ☆ Insect pests can be monitored through visual observation, pheromone and light traps, sticky traps, water traps, yellow traps, sweep nets, beating trays, and pitfall traps.
- ☆ In organic farming, biological agents can be imported to locations where they don't naturally occur, or farmers can make a supplemental release of natural enemies, boosting the naturally occurring population.

6.3.2 Predators that Prey in Fruit Crops

- Amoebae: Soilborne fungi, bacteria
- Anthocorid bugs: Spider mites, thrips, aphids, pear psylla, young scale, various insect eggs
- Bigeyed bugs: Lygus bugs, aphids, leafhoppers, spider mites
- Collembola: Fungi
- Ladybird beetles: Aphids, scale insects, pear psylla, mealybugs, other soft-bodied prey
- Lacewings: Aphids, scale insects, mealybugs, pear psylla leafhoppers, thrips, mites
- Mirid bugs: Spider mites, aphids, leafhoppers, pear psylla, scale insects
- Mycophagous mites: Fungi, *e.g.* grapevine powdery mildew
- Nematodes: Soilborne fungi, bacteria, other nematodes
- Predatory mites: Plant-feeding mites
- Spiders: Pear psylla, aphids, leafhoppers
- Syrphid flies or flower flies: Aphids, scale insects

6.3.3 Parasites Hosts in Fruit Crops

- Aphelinid wasps: Aphids
- Tachinid flies: Caterpillars, beetles
- Trichogramma wasps: Moth eggs
- Bacillus thuringiensis (bacterium): Butterfly/moth larvae
- Pseudomonas fluorescens (bacterium): Fungi
- Polyhedrosis virus: Butterfly/moth larvae
- Beauveria bassiana (fungus): Many insects
- Trichoderma harzianum (fungus): Pythium, Rhizoctonia and other pathogens
- Ampelomyces quisqualis (fungus): Powdery mildew
- Arthrobytris (nematode-trapping fungus): Nematodes
- Steinernema (nematode): Insect larvae
- Pasteuria penetrans (bacterium): Nematodes

6.4 Various Mechanical Control (Traps, Sticky Plates *etc.*)

6.4.1 Insect Trap

- It is used to monitor or reduce populations of insects/arthropods by trapping and killing them.

- ☆ It uses food, visual lures, chemical attractants and pheromones as bait and are installed so that they do not injure other animals or humans resulting into residues in foods/feeds.
- ☆ It varies in shape, size and construction and reflect the behavior or ecology of target species.
- ☆ It includes light traps, adhesive traps, flying insect traps, terrestrial arthropod traps and aquatic arthropod traps.

Different Insect Traps

Light Traps

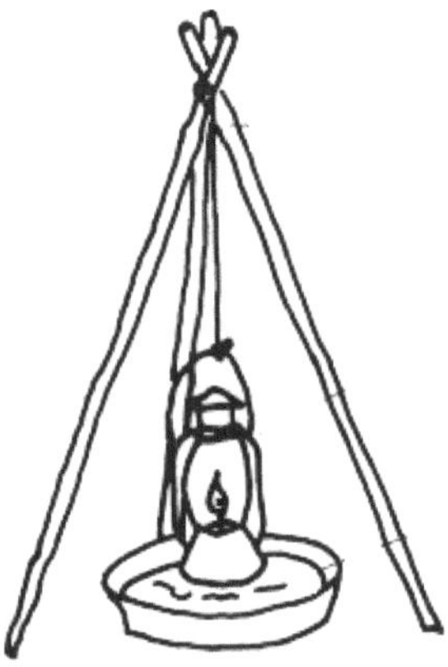

- ☆ These are set up at night which attracts different flying insects such as moths, mosquitoes, chafer beetles, American bollworms, army worms, cutworms, brown rice plant hopper, green rice leaf hopper, rice black bugs, rice gall midges, rice stem borers and tomato hornworms.
- ☆ It is normally constructed using bamboo/wooden poles and are arranged as tripod which are set downwards in ground to secure from wind/animals.
- ☆ In this, lantern is suspended from top over the bowl of water with less oil in it.
- ☆ The best time for placing light traps around garden/field depends on life cycle of insect and development stage of crop.
- ☆ The correct time of installation is soon after the moths emerge and before laying of eggs.

Fly Trap

- ☆ It is a large board that measures 30cm x 30cm which are painted with bright yellow/orange colors and is covered with adhesives such as oil or glue.
- ☆ There are different pests that are attracted to different colors as flies get attracted to bright color of board and flies on it which gets stuck in oil or glue and dies.

Pheromone Trap

- ☆ Pheromone is a sexual attractant which is produced by female insects.

- ☆ If trap is bait with this, it attracts male insects in trap from which they cannot escape.
- ☆ This trap is prepared by commercial companies and is costly to farmer/grower.

Adhesive Trap

- ☆ It is a sticky trap which has simple flat panels or enclosed structures, that are baited, that ensnare insects with adhesive substance.
- ☆ Bait less has nickname as blunder trap, where pests blunder in them while wandering or exploring.
- ☆ It is widely used in agricultural and indoor pest monitoring.
- ☆ Baited shelter trap like Roach Motel and similar enclosures have adhesive material to trap the insects.

Terrestrial Arthropod Traps

- ☆ Pitfall traps are used for ground-foraging and flightless arthropods like Carabid beetles and spiders.
- ☆ Pitfall traps has bucket or container in soil or other substrate so that its lip is flush with substrate.
- ☆ Grain probe trap is used to monitor pests of stored grain having long cylindrical tube with many holes along its length that can be inserted at many depths in the grains.
- ☆ Soil emergence traps has inverted cone or funnel with collecting jar on top which are used to capture insects with subterranean pupal stage.
- ☆ Emergence traps monitor important disease-vectors like Phlebotomine sandflies.

6.4.2 Sticky Plates

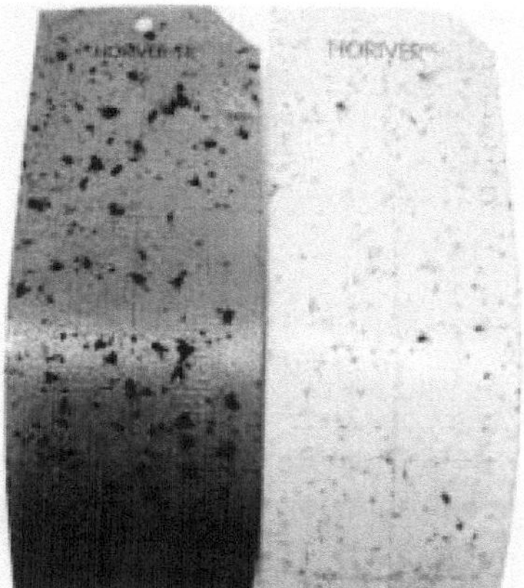

- ☆ Sticky plates are handy tools to monitor pest presence.
- ☆ These plates come in 3 colors, yellow, blue and black.
- ☆ Yellow color plate is especially suitable for monitoring white fly, aphid, leaf miner and sciarids
- ☆ Blue color plate is suitable to monitor thrips.
- ☆ Black color sticky plate is used to monitor the moth Tuta absoluta.
- ☆ Normally, sticky plates are available 25 x 10 cm, 25 x 20 cm and 25 x 40 cm sizes.
- ☆ To put sticky plates in greenhouse/fields, metal hooks are used to attach to greenhouse pillars using a magnet.

6.5 Advantages of Biological Control of Insects, Pest and Diseases

6.5.1 Introduction

- ☆ Biological control is a component of an integrated pest management strategy that shows the reduction of pest populations by natural enemies and involvement in active human roles.
- ☆ It is a method of controlling insect pests and diseases using other organisms that rely on predation, parasitism, and herbivore, or some other natural mechanisms with active farmer's management interaction.
- ☆ Biological methods are the use of beneficial organisms that can be used in the field to reduce insect pest populations.
- ☆ It is grouped as; importation or classical biological control which introduces pest's natural enemies to locations where they do not occur naturally, augmentation which involves the supplemental release of natural enemies, boosting of naturally occurring population and conservation, involving conservation of existing natural enemies in the environment.
- ☆ The role of beneficial species on pests is relatively more important in organic agriculture than in conventional agriculture, as organic growers do not have recourse to highly potent insecticides with which to tackle major pest problems.

6.5.2 Advantages

Biological control is controlling pests with their natural enemies such as parasites, predators, diseases and competing organisms.

- ☆ The biological control minimizes the environmental, legal and public safety concerns, Integrated pest control uses the bio-agents in combination with the other measures.

- It is a specific strategy where whatever the predator is introduced, will only control the targeted pest population, making it a green alternative to chemical or mechanical control methods.
- It allows the fruit to be left uninterrupted while the weeds are destroyed.
- It has no adverse effect on human health or on the environment.
- It is self-sustaining and cost-effective.
- In this, there is no need to find and identify every individual weed to be treated as the effective agent will search for suitable plants of weed.
- In this, the development of host resistance is not a problem and is compatible with other control techniques.
- It reduces the competitiveness and reproductive capacity of weed, making it more manageable.
- It does not use chemicals and uses many organisms which are either predators or parasites to pest.
- It can be applied and implemented anytime as it does not pollute the environment.
- In this, there is a restricted danger of damage to non-target plant species and does not create new problems.
- Its effectiveness is based on self-perpetuation and self-propagation, so on establishing a control agent in a specific area, it reduces target pest in acceptable threshold for a longer period.
- It is cost efficiency as yield benefit is less than yield achieved by agrochemicals, where the primary cost of the biological control agent is lower than chemical pesticides.
- It offers less risk of residues in the food chain and is suitable for the exotic pest which is not closely related to indigenous beneficial species.
- It is less expensive as compared to chemical control.

6.6 Biopesticides

6.6.1 Introduction

- Biopesticides are biological agents in the form of natural enemies or naturally occurring substances for the controlling pests.
- It includes microbial pesticides, mostly used-Bacillus thuringiensis, others are plant-Incorporated-Protectants (PIPs) and biochemical pesticides like insect sex pheromones.
- Biopesticides are characterized by having minimal or no risk to the environment, natural enemies and non-target organisms due to their mode of action, rapid degradation and small amounts applied to control pests.

- They are slow-acting, having relatively critical application times and suppress rather than eliminate the pest population.
- They have limited field persistence and shorter shelf life and present no residue problems, so they are approved for pest management in organic crops.

6.6.2 Advantages

- They are inherently less toxic than conventional pesticides.
- They affect only target pest and closely related organisms, in contrast to broad-spectrum conventional pesticides that affect organisms as different as birds, insects, and mammals including humans.
- They are effective in small quantities and decompose quickly, resulting in lower exposures.
- They avoid pollution problems which can be caused by conventional pesticides.

6.6.3 Examples

- Verticillium lecani is an entomopathogenic fungus that controls sucking pests like, aphids, jassids, mealybugs, whiteflies, mites, *etc.* Fungal hyphae enter in an insect body, parasitization and sporulation and kill insects in 5-7 days.
- Beauveria bassiana is an entomopathogenic fungus that can control all verities of borers, thrips, *etc.* The digestion of insect cuticle enzymatically by fungus and parasitization will control the insects in 5-7 days.
- Paecilomyces lilacinus is an entomopathogenic fungus used to control all types of nematodes. The fungal hyphae enter in an insect body, parasitization, and sporulation and kill the insects in 5-7 days.
- Beauveria bassiana is an entomopathogenic fungus that controls all borers, thrips, *etc.* The digesting of insect cuticle enzymatically by fungus and parasitization controls the insects in 5-7 days.
- Biofungicide on *Trichoderma viride* and *Harzianum* can be used to control soil-borne fungal diseases like root rot.
- Biofungicide based on *Pseudomonas fluorescens* can be used for controlling foliar diseases and acts as a growth promoter by secreting growth factors.
- *Trichoderma* and *Paecilomyces* in combination are used to control fungal pathogens and nematodes in the soil.

6.7 Record Keeping System

6.7.1 Records

Keeping track of what is happening on the farm/field requires some records. Good field management requires a good useful set of records that do not ensure success of farms, but success is unlikely without them.

Record-keeping is a task that has been stated many times by eager growers/ farmers, only to be dropped later. Reasons that complete, accurate records are not kept revolve around lack of time and motivation. Keeping records is productive work.

A good record-keeping system should take into account the growers needs, abilities and limitations. Organic crop growers should need to understand the purpose of keeping the records and when the purpose is known, it will help them in deciding which records are useful and need to be kept.

With accurate facts and figures is useful while borrowing money, seeking government support and completing tax returns. From reliable farm/field records, the equity proportion of total assets owned can be updated to assist with future farm investment programs.

Farm records can help wheat growers in evaluating wheat enterprise on mixed farm while farm advice will be more effective if the adviser knows exactly what is happening on the farm. Records and statements show what has occurred or is occurring on the farm while budgets predict what might happen in the future.

Records provide organic growers with data, information, and knowledge. There are four uses for farm records such as:

- ☆ Use as a service tool that covers income tax calculations, estate planning, business arrangement reconciliation and obtaining and managing credit.
- ☆ Providing data for financial analysis and diagnostic instruments like identifying the strengths and weaknesses of the business.
- ☆ Progress indicator where a series of records are required to monitor work progress.
- ☆ Forward planning tool for keeping past records to project future cash flows for current and modified farm management practices.

6.7.2 Details of Record

Records of the farm comprise of:

Farm Diary

- ☆ It is a form of the small notebook where the organic grower should record key facts and figures about farm business and its day-to-day activities.

- Farmers/wheat growers who use diary find important facts and figures which could be easily lost or forgotten are stored permanently in this for future reference.

Crop Records

It is useful to record what happens to each plot or crop each year that will include:

- Crop,
- Fertilizer applications,
- Agronomy (seeding, weeding, sprays),
- The visual impression of the crop,
- Harvest interval of forage crops
- Crop yield

Unit Costs

It covers all main farm inputs like:

- Fertilizers,
- Fuel,
- Irrigation water,
- Concentrates and their ingredients,
- Purchased forages,
- Stock purchases

The details are required for routine book-keeping and in monitoring seasonal changes and will help in planning future purchases.

Unit Returns

- It includes returns from all farm outputs such as the sale of the crop, manure, *etc.*

Crop Stock Accounting

- The main purpose of crop stock accounts is to monitor net losses and gains in income and to distinguish the increases and decreases due to changes in market value.
- When quantifying total stock value, wheat grower has to take into account both in change size and change in the unit of the crop.

Plant and Improvement Records

It covers records related to:

- Date and cost of purchase or installation,

- Annual depreciation,
- Insurance and registration,
- Fuel use,
- Hours use,
- Repairs and maintenance

Key Financial Records

- It covers details related to interest and principal repayment schedules to plan repayments.

Other Farm Charges

It includes records related to:

- Rates and government charges,
- Telephone and office upkeep charges,
- Laborer's wages,
- Capital investments,
- Unusual weather events,
- Important meetings,
- Credit repayments,
- Changes in the valuation of capital

Personal Expenses

- It includes expenses related to personal input wages which include the valuation of any farm products used for personal consumption.

Different Records

- It covers costs related to crops, fertilizer, seeds, spray, labor, water, machinery, harvesting, processing, transport and selling.
- In this, wages are broken down into different farm activities.

6.7.3 Importance of Records

- They are management tool which allows wheat growers to measure how efficiently they are using resources and to determine whether or not they are having an income.
- The records are essential for planning and decision making.
- The records are used in obtaining credit.
- The records are used for income tax.

6.7.4 Farm Accounting

It is the process of measuring and recording all farm resources and all farm transactions with financial consequences in a systematic way. The conduct of the farm accounting takes a large part of organic crop grower/farmer's time and effort which gets compensated in account of the following reasons:

- ☆ It permits the farmer/wheat grower to find out the size of income which is derived from the farm.
- ☆ It helps in knowing the total value of farm business and to know which part is actually owned by him/her and which by others.
- ☆ It will help in detecting loss or theft of cash or stock.
- ☆ It provides necessary data for correct income tax assessment.
- ☆ It helps in claiming expenses.
- ☆ It helps in keeping complete information as farmers/organic crop growers dislike paperwork since they are busy with their farm work.

6.7.5 Farm Financial Records

Farm transactions can be recorded when they occur or at the time of cash transactions. Since organic crop growers/farmers are involved in a lot of cash transition, so they should use accrual method of accounting, which will help them in the tracking of their money, inventory, sales, *etc.* For example, when recording sales of wheat for a specific time period, all the expenses needed to generate that wheat are relevant when calculating the cost of production.

Every physical item on-farm can be placed into the following categories:

1. Assets: It is required to generate income that can be either owned/purchased on credit that can be a combination of physical and monetary values, where physical quantity is multiplied by the unit price to have its value.
2. Liabilities: It relates to items on credit.
3. Equity: It serves as an asset with fewer liabilities.
4. Revenue: It is a source of income that is generated from the business.
5. Expenses: It involves the costs that are required for a business to generate revenue.

Annual financial accounting comprises of:

1. Opening balance sheet: record all assets and liabilities at the beginning of the financial period.
2. Closing balance sheet: record all assets and liabilities at the end of the financial period.
3. Cash flow: record all cash inflows and outflows in business.

4. Income statement: calculate wheat enterprise profit/loss before or after income tax.

6.8 Safety Measures and First Aid

6.8.1 Risk at Farms

Risk is a chance that someone could be harmed by hazards like working from ladders or working around electricity. A risk assessment is an important step in protecting people, business and complying with applicable laws.

Farmers and crop growers are particularly exposed to ergonomic hazards such as:

- ☆ Tiring or painful positions
- ☆ Carrying or moving heavy loads
- ☆ Standing or walking and repetitive hand or arm movements
- ☆ Work-related musculoskeletal problems

Farmers live with risk and make decisions every day that affect their farming operations. Many factors that affect the decisions which farmers make could not be predicted accurately such as:

- ☆ Change in weather conditions
- ☆ Drop in prices at the time of harvest
- ☆ Non availability of hired labor during working times
- ☆ Breakage of machinery and equipment when required
- ☆ Dying of draught animals
- ☆ Change in government policies

The above changes are examples of risks which farmers face in managing their farm business. Such risks affect farm profitability.

6.8.2 Sources of Risk

The most common sources of risk in farming can be in areas of:

- ☆ Production: Production risk occurs from uncertainty of factors that affects quantity and quality of farm produce.
- ☆ Marketing: Marketing risk occurs due to variability of product prices and uncertainty of future market prices.
- ☆ Financial: Financial risk occurs when money is borrowed to finance the farm business.
- ☆ Institutional: Institutional risk occurs due to unpredictable changes in provision of services.

- Human: Human risk involves risk to farm business that is caused by human illness and personal situation of farmer's family.

Based on farmer's attitude, risk divides farmers among:

Risk-averse Farmers

- The risk-averse farmers avoid taking risks.
- They tend to be more cautious individuals with preferences for less risky sources of income.
- They sacrifice certain amount of income to reduce the chance of low income and losses.
- They does not refuse to accept any risk at all.
- The risk averse farmer seek to be compensated for the risk taken by receiving a higher return than would normally be obtained if there were no risk.

Risk-taker Farmers

- The risk-taker farmers are people who are open to more risky business options.
- They choose the alternatives which gives some chance of higher outcome, even though they may have to accept lower outcome.
- They are influenced by the return they could receive.

Risk-neutral Farmers

- Risk-neutral lies between the risk-averse and risk-taking positions.

6.8.3 Workplace Risk

The main causes of risk/accidents due to equipments while working at site include:

- Transport (running over or vehicle over turnings);
- Falling from height (from fragile roofs, trees *etc.*);
- Struck by moving or falling objects (bales, trees *etc.*);
- Trapped by something collapsing or overturning;
- Livestock related accidents and fatalities;
- Asphyxiation/drowning.

Workplace Risk due to Farm Vehicles

Farm vehicles and implements are involved in many farm injuries and fatalities, so number of factors that will contribute to risk, will include:

- Unrestrained passengers being carried in the tray or on trailers

- ☆ Unsupervised children around moving vehicles
- ☆ Inexperienced drivers
- ☆ Poorly maintained vehicles.

Vehicle operator should always ensure that:

- ☆ Passengers use seatbelts
- ☆ The driver is trained and licensed
- ☆ Adopt safe driving methods
- ☆ Passengers are not carried in the back/tray
- ☆ The vehicle is maintained regularly
- ☆ Children are kept away from driveways and moving vehicles.

Workplace Risk Due to Harvester

The hazards due to harvester include:

- ☆ Entangled with levelling or discharge of augers in grain tank
- ☆ Falling from machine during pre/post season cleaning
- ☆ Contacting overhead power lines
- ☆ Being running over
- ☆ Contacting with knife, reel, stripper rotor, straw, chopper or spreader
- ☆ Being trapped under header or injured by header falling from its transport trailer

To avoid risk don't:

- ☆ Climb or reach in grain tank unless the engine and all augers are stopped, and the ignition key has been removed.
- ☆ Climb around areas of the combine unless there is purpose-built access available.
- ☆ Carry passengers on the header unless passenger seat is provided.
- ☆ Allow people to jump on or off from the harvester when it is moving.
- ☆ Work under an unsupported header
- ☆ Carry out maintenance or try to adjust automatic sensors with the engine running.
- ☆ The park, carry out maintenance or extend the unloading auger when under or near overhead power lines
- ☆ Run the combine with guards raised or removed.
- ☆ Allow children on or near the combine.
- ☆ Brake harshly, especially on slopes.

Harvester operator should:

- ☆ Make sure that anyone working on the harvester has read and understood the instruction book
- ☆ Carry out a 'Safe stop's procedure before carrying out maintenance.
- ☆ Have a system for thorough preseason maintenance checks to ensure that all brakes, guards and other safety devices are ineffective working order.
- ☆ Use any grain sampling mechanism provided.
- ☆ Have a safe system of work for combining paddocks where there are overhead power lines.
- ☆ Use appropriate farm maps as reminders when daily work is being planned.
- ☆ Agree on routes for travel with a local company that takes into account safe clearance distances
- ☆ Follow correct procedures for working under the header or when transferring header on and off its transport trailer.
- ☆ Provide and use correct tools and knife guard to aid safe removal, handling, and storage of the cutting knives.
- ☆ Work with cab door shut
- ☆ Be particularly careful when reversing
- ☆ Wear slip-resistant footwear and avoid wearing loose clothing.
- ☆ Take care on steep ground and avoid sudden changes of direction.
- ☆ Park across slopes if possible and ensure the handbrake is fully applied before leaving the vehicle
- ☆ Make sure all guards are in position and correctly fitted before starting work.
- ☆ Regularly clean straw and chaff deposits from the engine compartment and around belts and pulleys
- ☆ Make sure there is suitable fire-fighting equipment available on the combine

General Safety

- ☆ Make sure all guards and shields are in place.
- ☆ Ensure all workers are trained in proper operating techniques and safety procedures when using machinery.
- ☆ Stop all moving parts before servicing the machine.
- ☆ Operators should take regular breaks.
- ☆ Protective footwear and close-fitting clothes are essential when working in and around machinery.

- ☆ Use suitable access equipment and take precautions for work at height during cleaning and maintenance operations.
- ☆ Be patient when working in difficult crop conditions such as laid crops or unfamiliar fibrous crops like linseed.
- ☆ Don't operate the machine beyond its capacity.
- ☆ Use your instruments and watch and listen for potential blockages – avoiding blockages is easier than clearing them.
- ☆ Use drive reversing mechanisms, when fitted, to the wind out blockages.
- ☆ Remember the baler operator – try avoiding leaving heaps in the swath.
- ☆ Remember the hazards posed by straw choppers and spreaders – allow adequate rundown time before approaching the rear of the combine.
- ☆ Combine fitted with the yield measuring meters using an ionizing radiation source need additional safety procedures that must be laid down, clearly understood and implemented.
- ☆ Exposure to high levels of grain dust causes ill health, including occupational asthma, farmer's lung, grain fever, chronic bronchitis, allergic eye, and nasal infections.
- ☆ Make sure the cab filter is of the type recommended by the manufacturer and is fitted and maintained in accordance with the manufacturer's instructions.
- ☆ When unloading on the move consider the other driver and avoid sudden maneuvers.

6.8.4 Managing Risk

Reducing risk is often used in combination with one another as no single strategy can cover all of the risks likely to be encountered. Farmers need to consider the risks simultaneously and to develop an integrated approach for better management. They need to recognize the advantages and disadvantages of each risk management option both individually and in combination. Each farmer should select an appropriate strategy based on their goals, attitudes towards risk and their personal and financial situations.

Farmers should choose and combine strategies based on their goals, attitudes towards risk and their personal and financial situations. Every farmer's family required finding their own ways of coping with risk. Farmers' responses to risk are as diverse as the risks that affect their farms.

Farmers can reduce risk by learning about and applying new technologies and practices designed to address specific risks common to their area of products like new varieties of seed are being developed and livestock is being bred with certain characteristics like:

- The drought-resistant seed for wheat;
- The bird-resistant seed for sorghum;
- Disease and pest-resistant seed species;
- Disease-resistant livestock species;
- Livestock bred to provide better productivity;
- Irrigation for high-value crops;
- Crops and livestock bred specifically to improve marketability.

Factors Involved in Managing Risk

Avoiding Production Risk

- Selecting low-risk activities is a way to reduce production risk where farmers choose reliability over potential profitability.
- In this, the farmer may forego an enterprise that has a high potential for income but also carries a high risk for loss and choose instead an enterprise which is less profitable but also less risky.

Using Flexible Farming

- Farming system flexibility is an important strategy for managing risk.
- A flexible farming system makes it possible for a farmer to make quick or short-term changes in production and sales.
- Farmers who sell cash crops also reduce risk by using available funds to enable them to change to another enterprise if the price of the main cash crop falls.
- By keeping farm systems flexible, farmers are able to make decisions in response to changing circumstances.
- If farmers work on general production plans, they should keep their options as open as possible to respond to opportunities and risks as they occur.

Production Diversification

- Diversification spreads risk and is a successful risk management strategy as not all farm enterprises and operations are likely to be affected in the same way by changing situations. It involves:
- Managing multiple farm enterprises together at any one time (or in the same season);
- Engaging in the same farm enterprise in different physical locations;
- Engaging in the same farm enterprise over successive periods of time (or seasons);
- Generating income from off-farm activities.

Engaging in the same enterprise in different physical locations

- Such type of risk managing strategy works on understanding that the same crops grown in different areas will not meet the same fate.
- Perhaps a crop grown in the area surrounding the house is more likely to be infested by rodents than the same crop grown elsewhere.

6.8.5 First Aid

Accidents on farms and ranches can be quite severe, and space in a first aid kit is limited, so it is important to choose items for kits wisely. Most farms and ranches require multiple first aid kits due to the many types of jobs and the dispersed areas of work in a production agriculture operation. Not only is it important to have appropriate first aid kits on your farm or ranch, but it is also important that you and others in your operation understand basic first aid and cardiopulmonary resuscitation (CPR).

Items in the First Aid Kit

Injuries vary from job to job in production agriculture, so first aid kits should be tailored to the potential injury that could result from a particular job. Immediately and properly examining and treating injuries save life and are essential to reduce pain and help injured people make a quick recovery.

The following items should be there in the first aid kit which includes:

- Sterile first aid dressings of different sizes
- Trauma dressings for covering large areas
- Small, sterile adhesive compresses in sealed envelopes
- Roller bandages with different sized cling bandages
- Rolls of adhesive tape in assorted widths
- Triangle bandages to use as slings/coverings over large dressings
- Antiseptic wash
- Tongue depressors
- Bandage scissors and heavy-duty scissors to cut cloth
- Tweezers to remove insect stingers or small splinters
- Splints of the required size for splinting broken arms and legs
- Sterile saline solution
- Safety pins
- Chemical ice bags to reduce swelling
- A pocket mask for resuscitation
- Disposable rubber gloves

- Eye goggles
- Emergency blanket

It is the duty of the medical supervisor/farm manager to check the first aid kit on a regular basis. Medications expire, so you may need replacing old ones. Organic crop grower needs to replenish supplies that he/she has used in the past, so make sure that more common treatments such as medications, bandages, *etc.* are always available. If possible use antiseptic solution to clean the scissors and injections.

6.9 Handling Tools and Equipment

6.9.1 Checking the Machine before Use

Before working with any machinery, farmer/operator should carry out the basic check to make sure that the machine/vehicle what he/she will be using is in the good working order and is safe for use. For this, the machine/vehicle should be checked with basic checking like:

- Mechanical defects – pay particular attention to items like brakes, wheels and tires.
- Guards and other protective devices should be correctly fitted and maintained in good condition.
- Check that Guards and other protective devices are in correct place, securely attached and working properly.
- The damaged or defective guards/devices should be repaired or replaced before the machine is used.
- Check that the stopping devices are functioning correctly, such as emergency stops.
- See that all controls should be clearly marked to show what they do.
- If workers are to be carried on the machine ensure that they can do so safely, *e.g.* check for safe means of access and that working platforms have guard rails *etc.*
- Hitching and attachment points – check that mounted, semi-mounted or trailed machinery are safely attached to the towing vehicle such as a tractor.
- Examine the conditions of drawbar/pick-up hitch, hitch rings, pins, clips *etc.*
- Carry out any pre-use checks as specified in the operator's manual.
- For self-propelled machines, make sure that mirrors are clean and properly adjusted.
- Check any other reversing aids are working.

Chapter 7

Harvest and Postharvest Management under Organic Farming

7.1 Harvesting Based on Demand of Type in Market

7.1.1 Harvesting

Harvesting is the operation of cutting, picking, plucking digging or combination of operations for removing the crop from under/above ground and removing the useful part of fruits from plants.

It can be carried out using:

- ☆ Slicing action with sharp tool.
- ☆ Tearing action with rough serrated edge.
- ☆ High velocity single element impacting with sharp/dull edge.
- ☆ Two elements scissors type action.

7.1.2 Tools for Land Work/Harvesting

Kudali

It is an important tool used for digging of soil, pits or basic digging work before preparation of nursery beds. It has a metal blade attached to wooden handle and is operated manually.

Garden Fork

It is used for breaking of soil clods and separation of medium size soil during preparation of nursery bed or fields.

Weeding Fork

It consists of long handle with a blade of handle teeth used for loosens the soil and helps in weeding. It is drawn manually using the handle to collect the weeds and cops of plants *etc.*

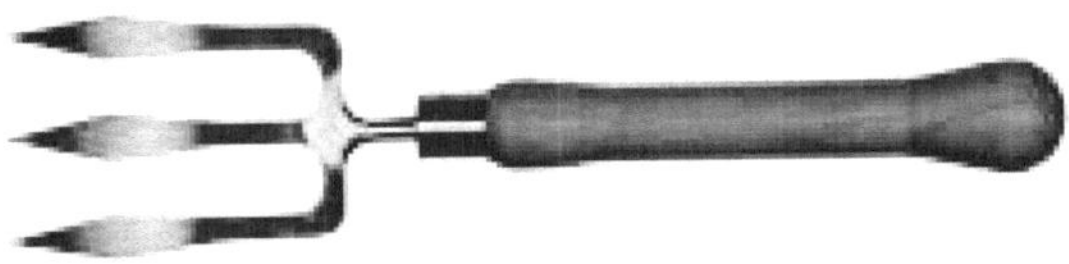

Crow Bar

It is an iron rod with one end pointed and other as a wedge shaped. It is used for digging out large or hand boulders from the soil and digging of pits. It is also used for breaking hard soil pan.

Shovel

It has an iron blade of spoon shape and wooden handle and is used in field transportation of dug soil for levelling of fields for preparation of nursery beds.

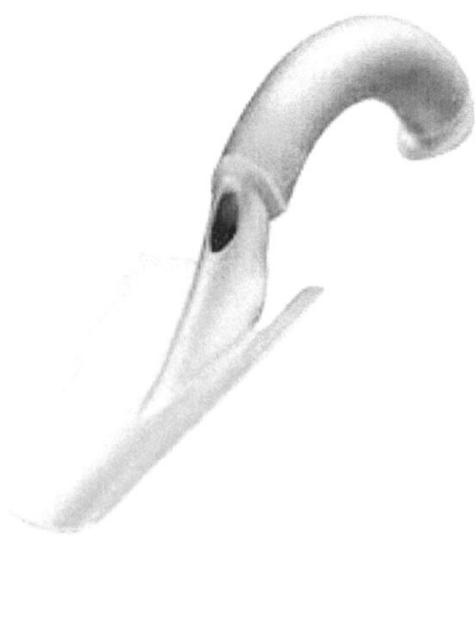

Pick-Axe

It is made of carbon steel having two edges with provision of axial hole for attachment with handle where one edge of pick axe is pointed while other edge is broadened. It is used for digging hard, compact and stony soils.

Spade

It is used for lifting and turning the soil and also for digging the pit, preparing channel for irrigation and drainage lines.

Hoe-cum-Rake

It is rectangular shaped metal blade with fork like fore edge used for digging, hoeing, earthing, levelling and collecting weeds.

Furrow Opener

It is used for opening narrow and shallow furrow after sowing seeds in nursery.

Hand Leveler

It consists of narrow rectangular metal blade attached to long wooden handle.

Trowel (Khurpi)

It is of shovel shape but small in size with iron blade and wooden handle. It can be made in many shapes as per local designs and requirements. It is used for hoeing, weeding and nursery plants and also for transplanting seedlings.

Axe

It has an iron blade which is fastened to wooden handle and is mainly used for felling trees and cutting branches.

Bill Hook (Darat)

It is made of iron curved at the far end which is used for cutting hardy branches of plant and other woody shrubs in the field.

Sickle

It is used for cutting grass and leafy vegetables.

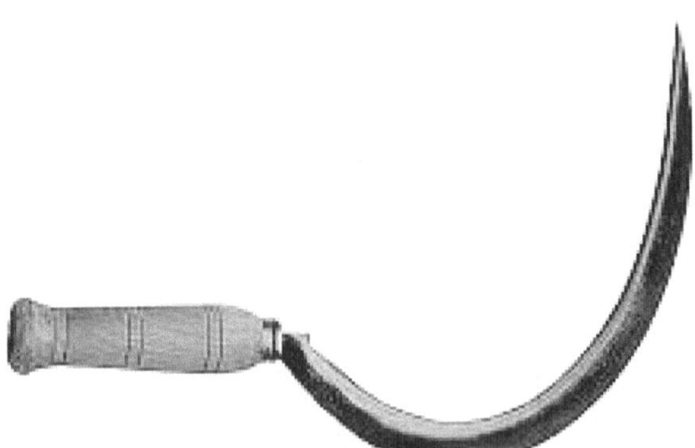

Wheel-Barrow

It is manually operated small trolley used for carrying nursery plants, compost, fertilizes, leaf litter, horticultural produce, stones *etc.* from one place to another place.

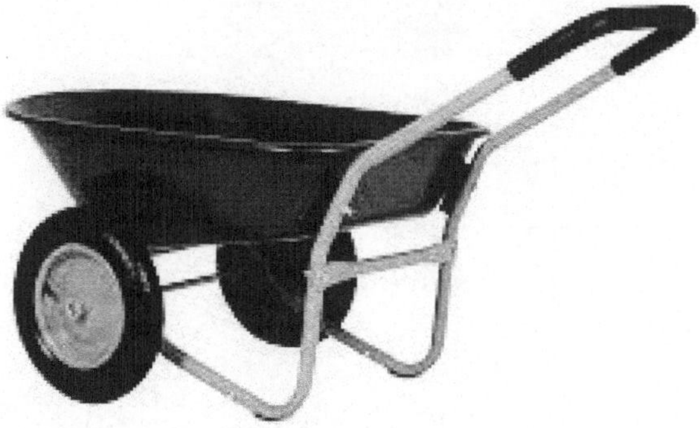

Cultivator

It can be tractor or hand driven implement used for tilling the soil efficiently. It has tynes of quality carbon comprising of different sizes depending upon the plough depth.

Hand Driven Implement.

Disc Harrow

It is used for pulverizing the soil and for turning crop residues, weeds and other debris into the soil. Also it can be used for deep ploughing and turning of soil.

7.1.3 Harvesting by Machines

Machineries are required at every stage of crop production, starting from seed bed preparation, pit digging, nursery raising and transplanting, seed sowing, weeding, intercultural operations, spraying and harvesting as also to meet the specific requirements of small and marginal growers.

Root Crop Harvester

- It is used for digging different crops like onion and garlic on single bed of 1.1 m wide.
- The machine has a digger blade with its width as 1144 mm and thickness 16 mm.
- The blade of harvester was mounted on machine at an angle of 20° with horizontal.

- ☆ It has an elevator conveyor which is attached behind the blade.
- ☆ In this, the spacing between MS rods of elevator conveyor is around 20 mm with slope of elevator conveyor as 18°.
- ☆ The conveying system has two oval agitators which are used to separate soil particles from tubers.
- ☆ In this, power to elevator conveyor is given by gear box.
- ☆ The blade has two coulter discs at outer ends for easy slicing and lifting of soil.
- ☆ The extension at rear is given for increasing separation time of soil.
- ☆ The triangular blade of 1730 x 60 mm dimension is used for digging of onion using two wheels for depth control.
- ☆ The harvester has a field capacity of 0.20, 0.26, 0.22 and 0.24 ha/h for digging onion, garlic and other bulb crops.

Tractor Mounted Vertical Conveyor Reaper

- ☆ It comprises of a reciprocating cutter bar assembly, seven crop row dividers, two vertical conveyor belts fitted with lugs, pressures springs, pulleys and gearbox for power transmission system.

- ☆ In this, crop row dividers are fitted in the front of cutter bar assembly.
- ☆ The star wheels are mounted over the crop row dividers.
- ☆ The machine is kept in front of the tractor where the power is given from the tractor PTO using intermediate shaft that runs below the chassis of tractor and coupling shaft.
- ☆ Here, the height of machine which is above the ground is controlled by tractor hydraulic using pulleys and steel ropes.
- ☆ Once the crop is cut by cutter bar, it is held in vertical position and delivered to one side of machine by lugged belt conveyors which falls on ground as windrow perpendicular to machine movement direction.

Tractor Operated Garlic Harvester

- ☆ In garlic harvesting, more labor is consumed for harvesting.
- ☆ Crop growers insist for single pass of duck foot cultivator for loosing of garlic bulbs from soil which are collected manually.
- ☆ In tractor operated garlic harvester/windrower, there is a triangular point knife blade which is mounted on MS flat for use in heavy soil.
- ☆ The harvester has chain type separating mechanism having MS bars of 12 mm at spacing of 38 mm along horizontal vibrating forks of 250 mm length at spacing of 200 mm for good separation of soil from clods.
- ☆ In this, windrower unit has MS bars at spacing of 25 mm which is kept at 900 to conveyor direction in unequal halves.
- ☆ In this, power is delivered from tractor pot to gearbox in 1:1 ratio.
- ☆ The gearbox with crank pinion arrangement gives power to oscillating vibrating fork.
- ☆ Tractor operated garlic harvester has an effective field capacity of around 0.26 ha/h.

7.1.4 Market Demand

Marketing is an important factor in determining the success of organic crops or related products which includes operations and decisions made by producer/grower that determine most of the marketable crops for production, delivery quality, produce, *etc.* Marketing does not begin after the crop is produced, but as an alternative needs to be considered even before production takes place.

It involves assembling, handling, storage, transport, processing, wholesaling, retailing and export of commodities with supporting services like market information, the establishment of grades and standards, commodity trade, financing and price risk management. India is bestowed with a lot of potentials to produce all varieties of organic products due to its various agro-climatic regions.

Features

- Many crop growers sell value-added products with high markups online.
- The current market for crops produced in fields includes general public, contractors, independent or chain retailers, other production nurseries, brokers, government agencies or a combination of these.
- Production nurseries that sell to the general public may do so via an affiliated retail nursery or by mail order.
- Production nurseries selling to retail operations may further specialize in sales to independent retail centers, chain retail centers or big stores.
- Sales to retail centers call for a more diverse group of bulb crop species and higher quality, higher-priced products in a greater selection of cultivars and sizes.
- Such market type usually involves many different customers in various locations needing different products at different times.
- A production nursery servicing retail centers has its own trucks for delivering crops.

Marketing System

- Marketing channels show how various market participants are organized to do the movement of product from producer to final consumer which is classified as per channel levels where the number of channel levels differs by product and rural and urban location.
- The marketable categories of onion comprise of spring, summer-fresh market onions and fall/winter storage onions.
- Fresh market or green, onions are pulled while the tops are still green and usually before a large bulb has formed.
- Fresh market onions can be identified by thin, light-colored skin and may have edible green tops.

- Fresh market and storage onion markets account for the largest share of onion use, other markets also account for significant production.
- Bulb crop moves from farmers to ultimate consumers through a number of marketing channels such as pharia, wholesaler, aratdar and retailer.

Channel -1	Farmers – *pharia* – wholesaler – *aratdar* - retailer - consumer
Channel -2	Farmers – *pharia* - retailer – consumer
Channel -3	Farmers – wholesaler – *aratdar* - retailer – consumer
Channel -4	Farmers - retailer – consumer

- In this, 55 per cent of produced bulb crop moves from farmers to consumer by Channel- 1 while 22 per cent moves through Channel-3. Also, 10 per cent and 13 per cent of bulb crop moves through Channel-2 and Channel-4.

7.2 Ideal Time of Harvest

7.2.1 Climatic

- Climate influences all components of crop production like cropping area and cropping intensity.
- Climate variability and change with factors like demand, price, policy and technology influences such components of production.
- It is seen that when weather is fatal to crops results in crop growth period, then an impact of fatal event is relevant to that of growing-season where climate varies in crop production in particular time.
- If no fatal weather event takes place, then growing-season where climate shows variation in crop production is seen in progress of crops.
- If the climate in an area is too wet or too dry, it will affect the cropping intensity of the crop.

7.2.2 Climatic Conditions

It includes:

- Temperatures
- Rainfall
- Relative humidity
- Average hours of sunlight,
- Other site-specific factors like frequency of extreme weather conditions

7.2.3 Timing

- ☆ Climate will determine what crops can be grown and at what time of the year as certain crops are frost sensitive while others have heat requirement or minimum soil temperature for its seed germination.
- ☆ It is important for crop grower to know when to harvest as correct time of harvest of crop is important in preventing crop losses.
- ☆ Timely harvesting of crop ensures good crop quality and market value.
- ☆ Crop grower should also known that harvesting the crop too early or too late can result in more unfiled or immaturity/prematurity of fruits which results in lower yield.
- ☆ Crop grower can easily determine the right time for harvesting of crop by analyzing moisture content of grains, sugar and nutrient content of fruits, visual properties of mature fruits, counting vegetation season days characteristic for each variety, *etc.*

The harvesting of various crops is done in warm/hot climate which includes:

Low Altitude Region	***High Altitude Region***	***Medium Altitude Region***
Wheat		
Button mushroom		
Boro Paddy	Spring potato	
	Summer potato	
Kharif Maize	Kharif Maize	Kharif Maize
	French bean	
Apricot, plum, peach and other temperate fruits		
Lady finger, cucurbits, brinjal, chillies		
Jute		
Pineapple, mango, banana, jackfruit		
	Late varieties of pear	

The harvesting of various crops is done in cold climate which includes:

Low Altitude Region	***High Altitude Region***	***Medium Altitude Region***
Rabi maize	Rabi maize	
	Soyabean	
Mustard		
Potato		
Arecanut, black pepper, oranges banana, corombolla		
Jute, mesta		

Low Altitude Region	*High Altitude Region*	*Medium Altitude Region*
Tomato		
Chestnut, persimon		
	Cabbage knol-khol, chillies, radish, brinjal	
	Ginger, turmeric	Ginger, turmeric
	Kharif paddy	Kharif paddy
Lady's finger, bitter gourd		
	Winter potato	
Sali paddy		
	Cauliflower	
	French bean, lady's finger, bitter gourd	French bean, lady's finger, bitter gourd
	Root crops	

7.3 Harvest Organic Crop Based on Usage

- Organic farming creates more profit and yields healthier produce by feeding a rapidly growing world population.
- Organic agriculture occupies 1 per cent of global agricultural land, making it a relatively untapped resource for one of the greatest challenges facing humanity, producing enough food for a population that could reach 10 billion in consecutive years, without extensive deforestation and harm to wider environment.
- organic farming systems produce yields that average 10-20 per cent less than conventional agriculture, they are more profitable and environmentally friendly.
- Despite lower yields, organic agriculture is more profitable for farmers as consumers are willing to pay more where higher prices compensate farmers for preserving the quality of their land.

7.4 Proper Harvesting Methods

7.4.1 Organic Crop Harvesting

- Organic products are grown under a system of agriculture without the use of chemical fertilizers and pesticides with an environmentally and socially responsible approach.
- Such a method of farming works at the grass-root level that preserves reproductive and regenerative capacity of the soil, good plant nutrition,

and sound soil management produces nutritious food rich in vitality that has resistance to diseases.

- Harvesting can be done manually, using hands or knives and it can be done mechanically with the use of rippers, combine harvesters or other machines.
- For harvesting of crops, the farmer can use certain strategies to determine the right time for harvesting which covers moisture content of grains, sugar and nutrient content of fruits, visual properties of mature fruits, counting of vegetation season days, characteristics for each variety, *etc.*

7.4.2 Harvesting Methods

There are four harvesting systems that use farmers all over the world:

- Manual system: It includes the use of traditional tools like hand cutting, hand threshing, animals for trampling.
- Manual harvesting with machine threshing: It is the combination of the manual and machine power.
- Combine harvesting: It covers all harvesting operations using combine without manual power. It is an efficient tool, which results in the lowest losses, but is expensive and requires large field areas.

Combine Harvester

- It is a machine which harvests grain crops such as wheat, oats, rye, barley, corn (maize), sorghum, soybeans, flax (linseed), sunflowers, peas, beans, lentils and canola.
- It is a combination of harvesting–reaping, threshing and winnowing.

- ☆ It carries hydrostatic drives.
- ☆ In this, cutter bar runs the full length of header below the reel where its teeth gets open and close repeatedly to cut the crops from its base.
- ☆ Here, behind cutter bar, cut crops get fed towards the center by spinning augurs which travels up conveyor to processing mechanism in main part of combine.
- ☆ In this, unwanted chaff and stalks, passes along conveyors towards the back of the machine.
- ☆ Although the greatest application of combination is in harvesting the small grains, corn and soybeans, these are also used for wide variety of small-acreage or specialty crops.

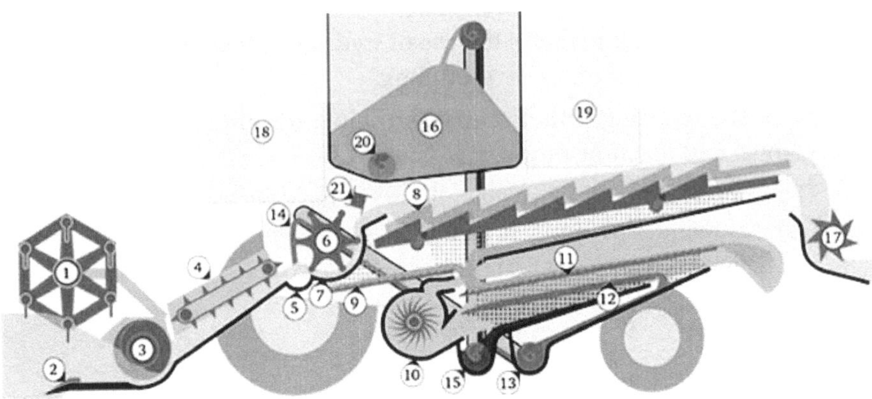

1) reel; 2) cutter bar; 3) header auger; 4) grain conveyor; 5) stone trap; 6) threshing drum; 7) concave; 8) straw walker; 9) grain pan; 10) fan; 11) top adjustable sieve; 12) bottom sieve; 13) tailings conveyor; 14) rethreshing of tailings; 15) grain auger; 16) grain tank; 17) straw chopper; 18) driver's cab; 19) engine; 20) unloading auger; 21) impeller

Its main functions include:

- ☆ Cutting the standing crops
- ☆ Feeding the cut crops to threshing unit
- ☆ Threshing the crops
- ☆ Cleaning the grains from straw
- ☆ Collecting the grains in a container

Tractor Mounted Vertical Conveyor Reaper

- ☆ It comprises of reciprocating cutter bar assembly, seven crop row dividers, two vertical conveyor belts fitted with lugs, pressures springs, pulleys and gearbox for the power transmission system.
- ☆ In this, crop row dividers are fitted in front of the cutter bar assembly.

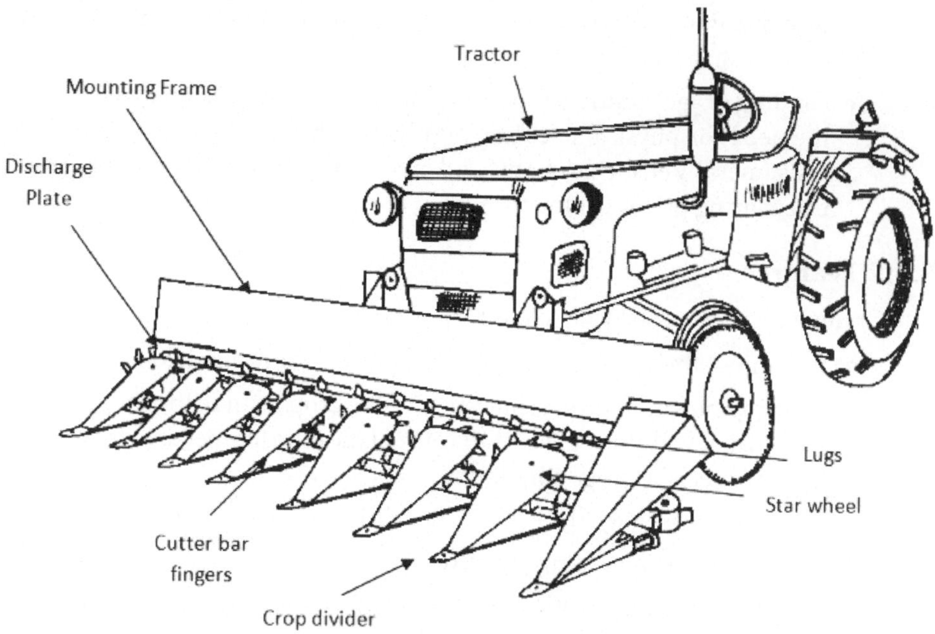

- The star wheels are mounted over the crop row dividers.
- The machine is kept in front of the tractor where the power is given from the tractor PTO using an intermediate shaft that runs below the chassis of the tractor and coupling shaft.
- Here, the height of the machine which is above the ground is controlled by tractor hydraulic using pulleys and steel ropes.
- Once the crop is cut by cutter bar, it is held in a vertical position and delivered to one side of machine by lugged belt conveyors that fall on the ground as windrow perpendicular to the machine movement direction.

7.5 Grading of Crop-Based on Size, Color, and Quality

7.5.1 Crop Grading

- Grades establish product quality requirements depending on food crops, which may include parameters like flavor, aroma, texture, color, moisture, size, maturity, firmness, shape, cleanliness, density, clarity, fat content, condition of crop and types of defects.
- Grading of crops/food involves the inspection, assessment, and sorting of various foods/crops in terms of quality, freshness, legal conformity and market value.
- The grading of crops can be done manually or by machines where in case of hand grading, crops/foods are assessed and are sorted while in case of

machine grading, crops/foods are graded by sorting the products based on size, shape, and quality.

- ☆ Grading of agricultural crops promote common trade language and avoid the need for physical checking and handling at multiple points, protect consumers by ensuring quality and protect the producer from exploitation by ensuring prices commensurate to the quality of produce.

7.5.2 Features

Grading of crops/foods will help in:

- ☆ Separating the crops into groups that meet certain standards.
- ☆ Showing potential processing value of the crop, its nutritional value and factors that will affect the overall value of the crop like need to dry or clean and for special storage conditions.
- ☆ Consumers to gauge the extra value of higher-quality standards.
- ☆ Facilitating the marketing of graded products.
- ☆ Improving the information related to flows and lower transaction costs.
- ☆ Ensuring that grains of equal value receive equal monetary compensation and that high-quality product receives the premium which allows customers to communicate their needs to producers.

7.5.3 Types of Grading

Grading is done to sort the, unlike lots of products into different lots as per the quality specifications. While grading, each lot should have similar characteristics in terms of quality of produce. With grading, products are divided into certain groups or lots as per predetermined standards.

Grading of product involves sorting as per size, quality and other characteristics which cover:

(i) Permissive/valuable grading: standards vary over time.

(ii) Contrived/de-centralized grading: in this, the packer sets his/her own lab approved grading lots by purpose.

(iii) Grading of producers level: In this, free grading services are given in market yards to crop growers for sorting their produce before offering in the market for sale.

Example

- ☆ Grading of apples is done as per fruit size and fruit appearance or quality.
- ☆ Based on fruit size, apples are graded manually in six grades.

- ✰ Based on fruit colour, shape, quality and appearance, apples are graded in three or more quality grades as AAA, AA and A; A,B,C; or extra fancy, fancy class I and fancy class II.

7.6 Packaging of Crop with Appropriate Material and Method

7.6.1 Introduction

Harvested crops are kept in packages, so as to protect it from spoilage, breakage, damage from external environmental conditions, pilfering and theft. Proper packaging of crops is important in reducing losses, especially in tropics as climate considerably increases the risks of deterioration. The main functions of packaging are to:

- ✰ Permit easy handling, whether manual or mechanical;
- ✰ Reduce product losses by theft;
- ✰ Protect product from external attack like humidity, insects, sunlight, *etc.*

There are various types of packaging for agricultural commodities, appropriate to product's characteristics and to the marketing system. The choice of the type of bag should also take into account not only its inherent toughness and resistance to humidity, sunlight and pests but also the type of handling anticipated.

7.6.2 Materials Used in Packaging

The packaging material for crops can be of:

- ✰ Biodegradable/compostable
- ✰ Paper
- ✰ Plastics and master batches
- ✰ Metals: steel, aluminum
- ✰ Glass

- ☆ Composites
- ☆ Others: jute, wood *etc.*

7.6.3 Package Selection/Design

- ☆ The design of the packaging should be such that it lead to safe arrival and effective promotion of crop/product.
- ☆ Package design should comprise of structural design, technical construction of package from functional point of view, graphic design, appearance of package and its promotional value.

The package so selected should be based on:

- ☆ Quality standards in terms of performance and presentation
- ☆ Type of crop
- ☆ Type of package quality (knitted bags, corrugated boxes, metal or plastic drums)
- ☆ Customer requirements
- ☆ Handling features
- ☆ Transporting capabilities

7.6.4 Packaging Range

The range of packaging can be subdivided into:

- ☆ Primary package is in direct contact with the food.
- ☆ For distribution, several primary packages are placed in secondary packages.
- ☆ Several secondary packages are placed in a unit load which forms tertiary package.

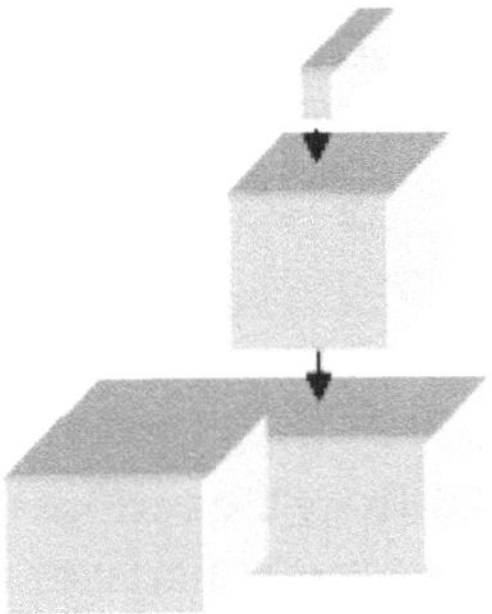

Packaging for Fresh Produce

Packaging serves as the main item of expense in produce marketing, so the selection of suitable containers for commercial marketing requires careful selection. Apart from the uniform-sized package for protection of produce, there are other requirements for a container, such that it should:

- ☆ Be easily transported when empties
- ☆ Occupy less space than when full.
- ☆ Be easy to assemble, fill and close either by hand or by a simple machine
- ☆ Provide ample ventilation for contents during transport and storage.
- ☆ Have suitable capacity as per market needs.

- ☆ Have suitable dimensions and design, so that it can be easily fit in available transport.
- ☆ Be cost-effective in terms of the market value of the commodity, for which it is used
- ☆ Be readily available from suppliers

7.6.5 Features

- ☆ Packing of crops should be small for easy handling during transit and as per market demand.
- ☆ Organic crops should be packed in jute/Hessian bags for transporting to a yard or brought as loose.
- ☆ For safe handling, 40 kg open mesh jute bags with 200-300 g weight should be used in the domestic market.
- ☆ For export, big crops are packed in 5-25 kg size open mesh jute bags.
- ☆ Nylon net bags, when used for packing have resulted in less storage loss due to good ventilation.
- ☆ Sacks and nets should be used for crop packaging for general-purpose jute sacks for agricultural commodities, open-weave sacks of sisal fiber, open-mesh nets of plastic materials and big bags to contain up to 1,000 kg.
- ☆ The rough nature of fiber gives sufficient stable weave and is similar to jute sacks which allow limited visibility of onions and impedance to less airflow.
- ☆ Open-mesh nets are used packages for different crops, which are red or orange in color.
- ☆ The slippery nature of plastics results in movement of threads allowing large holes to open up, so to overcome such problems, alternative nets are produced to give fully stable mesh and stronger bags which are degraded by sunlight and should not be left outdoors for long period before use.
- ☆ Crops should be stored as per required bag types and size of the crop, so crops should not get spoiled due to rubbing, ventilation, *etc.*

7.6.6 Proper Packaging of Crop Production in Clean and Disinfected Bags

- ☆ Foodgrains like wheat and paddy should be stored in special bags.
- ☆ After threshing, grains are separated from chaff or straw by lifting and tossing the threshed material so that lighter chaff and straw gets blown to one side while heavier seeds fall down vertically.

- Before crops are kept in bags for storage or processing, they need to be dried which can be done by spreading them out in the sun or with the artificial dryer.
- Crops after drying are stored in clean and disinfected bags for further processing and should be inspected regularly for signs of spoilage and moisture content.
- If the crops have moisture, it should be again dried.
- Once the crops are fully dried, they should be protected with insecticides and should be stored in rodent-proof bags/containers.

7.6.7 Precautions

- Packaging materials that affect the organic nature of contents should be avoided.
- The use of PVC materials should be prohibited.
- Laminates and aluminum should be avoided.
- Recycling and reusable systems shall be used wherever possible.
- Biodegradable packaging materials shall be used.

7.7 Ideal Storage Conditions (Temperature, Moisture, *etc.*)

A crop may need to be stored for several weeks or months, or even up to 11 months to meet market demand. Often, a better price for crops is offered months after harvest, so storage may also be an economic advantage. To maintain crop quality during storage, insects must be kept out and the crop must be stored at the proper temperature and moisture conditions.

7.7.1 Factors Affecting the Storage Life

The natural limits to the postharvest life of all types of fresh produce are severely affected by other biological and environmental conditions:

- Temperature. An increase in temperature will result in an increase in the rate of the natural breakdown of produce as produce reserves and water content become depleted, so the cooling of products will help in extending produce life by slowing the rate of breakdown.
- Water loss. High temperatures and injuries to produce will increase the loss of water from stored produce beyond that unavoidably lost from natural causes, so maximum storage life can be achieved by storing undamaged produce at low temperature as tolerable by crop.
- Mechanical damage. Damage caused during harvesting and subsequent handling will increase the rate of deterioration of produce and renders it a liability to attacks by decay organisms, so mechanical damage to root

crops causes heavy loss owing to bacterial decay which must be remedied by curing roots or tubers before storage.

- Decay in storage. The decay of fresh produce during storage is caused by infection of mechanical injuries. Many fruits/vegetables are attacked by decay organisms, which penetrate by natural openings or by intact skin, as such infections get established during the growth of the plant in the field but lie dormant till after harvest and often becomes visible only during storage or ripening.

7.7.2 Proper Moisture Content before Storage

- One of the most critical physiological factors in successful crop storage is moisture content.
- High moisture content leads to storage problems as it encourages fungally and insect problems, respiration and germination.
- Moisture content in growing crops is naturally high and only starts to decrease as the crop reaches maturity and grains are drying.
- In natural state, seeds have a period of dormancy and germinate when re-wetted by rain or by natural adequate moisture content.
- The moisture content of the wheat crop should be checked by using the formula; weight of moisture/Weight of wet sample x 100.
- Crops are harvested at a moisture content of 18–25 per cent (wb), although it can be substantially higher or lower depending on maturity, season, weather pattern and drying facilities.

Keeping Moisture Level Balanced

For keeping crop moisture level low/maintained, the following methods should be applied:

- Before storing, it should be properly dried.
- Storehouses should be well ventilated.
- Maintenance of adequate free space (1 1/2-2 ft.) between stacks.
- Stacks should be well separated from the wall and ceiling.
- Time to time aeration of the storehouse reduces moisture content and prevents pest growth.
- Damp and wet bags should not be used for storing.
- Crops should be frequently inspected at least once in a month during the dry season and once in a fortnight during the rainy season.
- The damp and moist grains should be disposed of at the earliest or should be removed from the storehouse to be dried in the sun.

7.7.3 Temperature Requirement for Proper Storage

- Crop grower is required to maintain the grain/crop temperature as per different regions and seasons.
- In cereal crops storage at correct moisture content and temperatures will tend to avoid pests and molds.
- There is a need for crop grower to measure the temperature of grain bulk as any sudden rise in temperature give an early and sensitive warning regarding the storage of crops.
- The temperature of the crop/grains should be maintained below 60°F to control insects and mold.
- The temperatures should be reduced to optimum storage level as early as possible in the following harvest, and crop/grain temperature should be managed by aeration of grains in fall, winter, and early spring.
- The aeration time required to achieve 60°F will vary due to airflow rates of equipment used and ambient temperatures.
- Aeration reduces grain moisture content from ¼ to ½ of a percent during one aeration cycle.

7.8 Market Rates of the Crop

7.8.1 Price Movement

Movement of products from producers to consumers involves costs, taxes, and cess which together contribute to the market price. Such prices vary with channels by which a particular crop/commodity passes through such as cost of packing, transport, weighing, loading, unloading, losses and spoilage.

Market price normally includes:

- Handling charges at a local point
- Assembling charges
- Transportation and storage charges
- Handling by wholesaler and retailer charges to customers
- Expenses on secondary service like financing, risk-taking, and market intelligence
- Profit margins took out by different agencies.
- Producer's share in consumer's rupee

7.8.2 Market Rate

The market rate of the particular crop is based on certain factors which include:

Market Margins and Price

To decide about crop pricing, crop grower should understand about:

- Price spread which shows the difference between the price paid by the consumer and price received by the producer for a certain quantity of crop production.
- What price spread comprises of such as marketing costs and margins of intermediaries to determine the overall effectiveness of the marketing system.
- Were highest margins are there for retailers in comparison with intermediates in secondary and terminal markets.
- Were the highest profits are there for the retailer in secondary and terminal markets.
- Profit as a percent of investment was maximum for the retailer.
- Reason for the highest profit and margins due to the high selling price of bulb crops.
- The market margin of the crop as calculated by deducting purchase price from the sale price and profit from the difference between the marketing margin and marketing cost.

Price Fluctuation

- Crop growers/farmers mostly have downside price risk in contrast to end-user who mostly have upside price risk.
- Managing price risk can occur through a spot transaction if the product is available or by forwarding transaction, if the crop/product is not available, either with physical forward contract or hedging price by the futures market.
- Selecting subjective price targets/triggers depends on the individual risk attitude and circumstance of decision-maker.
- An unrealistic target price adds the profit margin to the current price in the bearish market when the commodity market intrinsically ignores the operational costs of the farmer or end-user.
- Crop growers sometimes are forced to accept prices below their cost of production, whereas end users may sometimes be forced to buy above their break-even limit.
- The alternative for farmers is to withhold product from the supply chain either by hoarding or pooling, whereas the alternative for end-users is a switch to substitute products or alternate operating methods.

- To minimize price risk, farmers/crop growers should understand the meaning of Minimum Support Price that is a form of market intervention that insure agricultural producers against any sharp fall in prices.
- The minimum support prices are announced by the government at the start of the crop sowing season on the basis of recommendations of Commission for Agricultural Costs and Prices.
- Minimum support prices are guaranteed price for crop production from the government as its main objectives are to support the farmers from distress sales and to procure food grains for public distribution.
- In case the market price for a commodity falls below the announced minimum price due to bumper production and glut in the market, government agencies purchase the entire quantity offered by the farmers at the announced minimum price.

Analysis from Previous Year Crop Rate

- Market information provides hints to crop growers about comfortable pricing which signals and gives performance about the particular crops in domestic markets to international markets and vice versa.
- The study of marketing costs and margins about bulb crops serves in the evaluation of marketing performance for input and output markets and across the entire market or value chain.
- Farmers usually assess prices before selling bulb crop in the market.
- By tracing the market cost of the crop, an estimate of magnitude and quantity traded can be analyzed from official records.
- Results from the survey show that brokers are the main information source, which accounts for 85 per cent for farmers followed by 9 per cent, which is information obtained from other farmers.
- Farmers get market information from brokers who often have the power to set the price to farmers.
- Due to seasonality, the nature of production in the district, small-holder farmers are always uncertain, which attributes to rising and fall of the price in the market.
- During a shortage of bulb crops in the market selling price rises while during peak production of bulb crops, price falls on average due to excess supply and absence of linking crop to export market destination.

Chapter 8

Undertake Quality Assurance and Certification in Organic Farming

8.1 Systems of TPC and PGS

8.1.1 Organic Certification

- ☆ It is a procedure in which a certifying body assesses the farm or company and assures in.
- ☆ writing that it meets the requirements of organic standards.
- ☆ In this process, the inspector transmits his/her findings to certifying the body as a written report where the certifying body compares the results of the inspection with the requirements of organic standards.
- ☆ Further, the certifying committee decides whether certification may be granted or not.
- ☆ It is a certification process for producers of organic food and other organic agricultural products, so in general, any business directly involved in food production can be certified, including seed suppliers, farmers, food processors, retailers, and restaurants.
- ☆ The organic certification requirements vary from country to country which involves a set of production standards for growing, storage, processing, packaging, and shipping.

Purpose of Certification

- ☆ Certification is aimed at regulating and facilitating the sale of organic products to consumers.

- When small farmers sell their produce directly at farmers' markets, more and more consumers are purchasing organic food by traditional channels like supermarkets.
- It is intended to assure quality, prevent fraud and promote commerce.
- For organic producers, certification identifies suppliers of products approved for use in certified operations.
- For consumers, certified organic serves as product assurance.
- Organic certification addresses the growing worldwide demand for organic food.

Organic Certification Agencies

There are two types of certification agencies.

- Indian certification agencies
- International certification agencies

Steps for Certification

There are certain certification steps which crop grower should follow:

- Submission of application
- Cost Estimation
- Invoicing and payment of fees
- The signing of inspection contract
- Inspection and submission of inspection report to the certified department
- Issuance of the certification decision
- Compliance of certification decision and issuance of the certificate
- Issue of Transaction Certificates/Inspection Certificates for traded lots

8.1.2 Third-party Certification

- It is the most widely used model which is carried out by an independent body that is a precondition for gaining access to larger markets for organic products.
- It involves individual certification of farmers, processors, distributors, company farms, *etc.* and group certification like farmer cooperatives, peoples organization, *etc.*
- The certification depends on external auditors which inspect and assure production quality adheres to organic certification standards.
- It is a certification driven process where external certification agencies do annual reviews on organizations production, farming, and internal

control practices and checks on adherence of organization practices against certification standards.

- Such type of certification is accepted locally in India and abroad.
- The certification is mostly done for groups with 500 farmers each, where fields of 35-45 farmers are tested in every group.
- The certification is too costly for smallholders.
- In this certification, accreditation is provided by the private or public sector often by public authorities.

Steps

- Third-party Certification is done by a third party without a direct interest in the economic relationship between the supplier and buyer.
- Certifications bodies, whether private or public organizations, can be accredited by independent third-party accreditation bodies.
- It is a formal and documented procedure by which, the third-party assures organic production standards are followed.
- To certify a field, crop grower is required to follow certain activities which involve:
    - Information related to standards: Crop growers should study organic standards that contain detail about what is and is unallowed for every aspect of farming, such as storage, transport, and sale.
    - Compliance: Crop growers should know about field facilities and production methods that must comply with standards involving modification facilities, sourcing, and changing suppliers, *etc.*
    - Documentation: Crop grower should know about the type of documentation required like detail field history and current set-up with information about the results of soil and water tests.
    - Planning: Crop grower should have idea about written annual production plan which needs to be submitted showing detailing about seed to sale, seed sources, field and crop locations, fertilization and pest control activities, harvest methods, storage locations, *etc.*
    - Inspection: Crop grower should have an idea related to field inspection which normally involves with a physical tour, examination of records and oral interview.
    - Fee: Crop grower should have an idea about the annual inspection/ certification fees to be paid to certifying agencies as per the location.
    - Record-keeping: Crop grower should keep written, day-to-day farming and marketing records which should cover all related activities to be used for inspection.

8.1.3 Participatory Guarantee Systems

Introduction

- Participatory Guarantee System is a quality assurance initiative that is locally relevant, which emphasizes the participation of producers and consumers that operate outside the framework of third-party certification.
- It is a clearly documented system of quality assurance for organic farm products which results in a written certificate of the same.
- It is a process in which people in similar situations assess, inspect and verify the production practices of each other and take decisions on organic certification. The approach of such a system towards certification is non-hierarchical which uses less paperwork as compared to a third-party certification system.
- The certification system offers a complementary, low-cost, locally-based system of quality assurance with more emphasis on social control and knowledge building.
- The requirements for such a system include mechanisms to ensure that crop growers understand the organic standards they are committing to, participate in peer -reviews and making publicly recorded pledges or declarations to uphold the organic standard.
- The idea of such a system is to encourage local and direct relationships between crop growers and consumers.
- The system program is focused on and better suited to small-crop growers and direct markets, which brings many growers that wouldn't have considered third-party certification into a system of committed organic production.
- The system makes an excellent base for international system programs as many of basic structures are already in place.
- Such a system with third-party certification serves different markets and different operators, without any need for competition among them.
- The system requires more work and involvement from the crop growers and offer whole-farm certification by allowing crop grower to market all their products as organic even to local markets.
- The system leaves ownership of the certificate with crop growers which are not always the case with the international system.
- Such system procedures are more flexible as they tend to be more inclusive and appropriate for the local social context they serve.
- The system programs challenged by low literacy levels of their crop growers decided to use video records for growers applications and declaration statements rather than written statement.

- Such system crop growers receive a certificate that can be used to show their organic status.

Features

In this

- The group of growers who are interested in organic farming takes a pledge to follow guidelines as suggested for a particular system.
- Peer review committee has five members from the local group which are responsible to monitor and inspect the fields of every grower in the group. Here one of the members of the committee will act as conveyor and after entering all details in proforma, the committee discusses each case for taking a decision on recommendation for eligibility.
- The facilitating center will be responsible to compile information that is received from local groups in the region and will forward the same to system committee for issuing of certificates to eligible local groups.
- Crop growers who are following organic farming regularly for certain minimum time of three years on total land holdings as per system guidelines will only get such system certificates while others would be considered as under conversion till they satisfy particular criteria.

8.2 Organic Production Standards

- Standards for Organic Production are notified in National Programme for Organic Production under Foreign Trade Act, 1992.
- It covers standards related to crop production plan, conversion requirement, nutrient management, disease management, *etc.* for agricultural crops.
- These organic production standards and accreditation system have been recognized by European Commission and Switzerland for unprocessed plant products as per country standards.
- Organic standards are sets of requirements which shows type of practices to be considered for organic.

These standards address various aspects of organic products such as:

- General farm production requirements and conversion periods
- Crop production needs and requirements for collection of wild products
- Animal production requirements
- Processing and handling requirements
- Social justice requirements
- Labeling requirements

As per organic production standards:

- ☆ Land will have no prohibited substances applied for at least 3 years before crop harvesting.
- ☆ Use of genetic engineering, ionizing radiation and sewage sludge is prohibited.
- ☆ Soil fertility and crop nutrient will be managed by tillage and cultivation practices, crop rotations, cover crops, supplemented with animal and crop waste materials with synthetic materials.
- ☆ Preference shall be given to organic seeds usage and planting stock, but farmer can make use of non-organic seeds and planting stock under specified conditions.
- ☆ Crop pests, weeds and disease can be controlled by management practices like physical, mechanical and biological controls.
- ☆ If practices are not enough, then abiological, botanical or synthetic substances that are approved for usage should be used.

8.3 Procedures in TPC and PGS

8.3.1 Certification

- ☆ Certification procedures and accreditation mechanisms are tools in the quality assurance system to ensure transparency and compliance with standards and regulations that define organic agriculture.
- ☆ Certification procedures include inspection and certification.
- ☆ Inspection aims to verify and ensure that the production and handling of a product is carried out in line with certification standards.
- ☆ Certification procedures should make it possible to track the flow from primary production through entire food chain right till the final consumer.
- ☆ All operators in product chain as farmers, processors, manufacturers, exporters, importers, wholesalers and retailers must be certified as acting in compliance with organic standards and regulations.

Inspection

- ☆ Standard inspection procedures shall be followed by inspection and certification agencies.
- ☆ In this, the inspector shall have access to all relevant facilities, including accounts and other documentation of Licensed Operator.
- ☆ The policies and procedures for inspection shall be documented which includes:
  - ❖ The basis for assignment of inspectors.

- Grounds for objection to inspection by Licensed Operator.
- Instructions for inspection visits.
- Inspection methods and frequency.
- Inspection requirements.
- Sampling requirements.
- Instructions for the preparation of reports.

8.3.2 Procedure in PGS

- In the operation of PGS, stakeholders are involved in decision making and essential decisions about operation of the system itself.
- In addition to being involved in the mechanics of the certification system, stakeholders mostly producers are engaged in the structured ongoing learning process, which helps in improving their working.
- The certification system is based on a participatory approach where shared vision, transparency and trust exists.
- The system gives movement, national recognition, and institutional structure without affecting the spirit of the system.

PGS Certification Process at Local Group Level

The processing steps involved in PGS certification process at Local Group Level will cover:

- Making groups by willing crop growers
- Pledging to follow organic
- The signing of the pledge and an agreement
- Following standards
- Taking endorsement from other group
- Register with Regional Council
- Peer review
- Certification decision
- Submission of group summary sheet to regional center for approval

PGS Certification Process at Regional Council Level

The processing steps involved in PGS certification process at Regional Council Level will cover:

- Receiving of application of groups
- Checking and granting registration
- Providing all documents

- ☆ Participate in group meetings, surveillance, and peer review.
- ☆ Facilitating group in data uploading on the web site
- ☆ Receiving and checking of the summary sheet
- ☆ Approval/non-approval of the certification decision
- ☆ A regional center cannot decide on an individual farm and can approve or reject entire group decision.

PGS Certification process at Zonal Council Level

The processing steps involved in PGS certification process at Zonal Council Level will cover:

- ☆ Capacity building of Regional Centers and Groups
- ☆ Trainings and Surveillance
- ☆ Appellate authority for groups
- ☆ Assisting system secretariat in sample collection
- ☆ Online/physical monitoring of program operations
- ☆ Noninterference in functioning of Local group and Regional Centre
- ☆ In adverse reporting or residue test positive, it will be communicated to the regional center for necessary action

8.3.4 Procedure of TPC

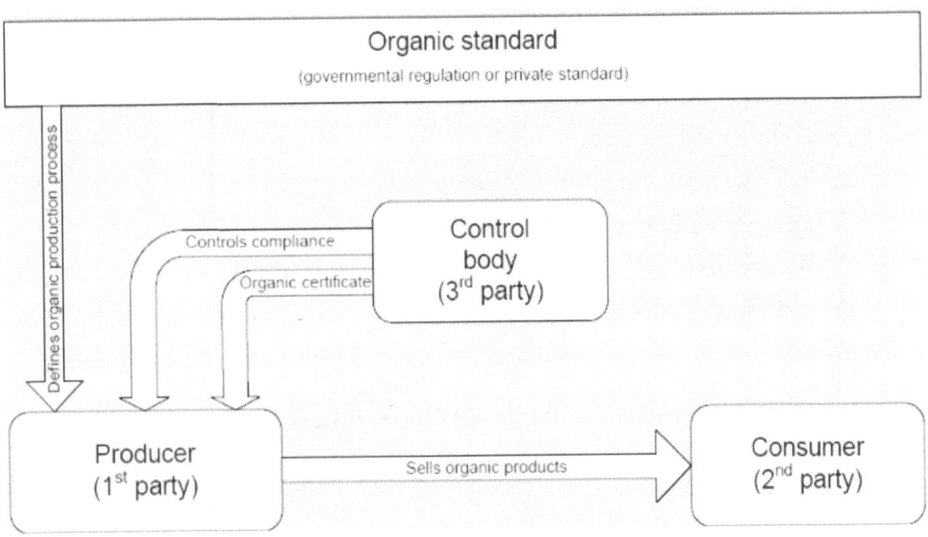

The process of getting a Third Party Certification involves following steps:

- ☆ The operator who can be producer, importer, and supplier applies to a particular third-party certifier body for certification.

- In such a case, third-party certification body will conduct a pre-assessment and documentation review of the operator's facilities and production operations.
- The Control body will conduct field audits/inspection and the inspector fills the inspection report.
- On confirmation and verification of the report, certifier issues a certificate, which allows the supplier to label its products as certified.
- The suppliers are responsible to pay for audit costs, but subsidies are paid to certain farmers.
- In this, it is important that every crop grower, processor, and importer in the organic farming supply chain should inspect at least once in a year.
- Further, certification needs to be renewed on an annual basis which covers all processing and trading steps involved for the sale of the product.

8.4 Benefits and Limitations of TCP and PGS

8.4.1 Third-party Certification

- A third-party certification system is in place to nurture quality innovation and consumer confidence in the marketplace.
- The certification system relies on external auditors to inspect and assure the quality of production adheres to organic certification standards.
- Third-party certification is a process-driven certification which is overhead costs for small and marginal farmers.

Benefits

- Such type of certification system is accepted locally in India and abroad.
- The system conveys impartial judgment as per the level of secure specification which allows purchasing a product that is at par.
- The certification system is compliant with applicable electrical code which assures proper installation and application of a product which reduces the risk of potential user injury.
- The mark of third-party certification shows a higher degree of reliability, which consequently increases the quality of the product.
- Due to the independent lab for testing, the time taken by such a system to complete the manufacturing process is shorter which leads to fast credible products.

8.4.2 PGS Certification

- The certification system is only for farmers or communities that can organize and perform as a group in the village or in close-by villages with continuous territory.

- ☆ The system is applicable to, on-farm activities which comprise of crop production, processing, and livestock rearing and off-farm processing by PGS crop growers of direct products.

Benefit

- ☆ It emphasis direct sales to consumers that results in low costs with high profits
- ☆ It leads to greater education about organic production techniques
- ☆ It gives a strong feeling of community and increases the confidence in the organic quality of products.
- ☆ It gives a good feeling about supporting local farmers.

Limitations

- ☆ It is only for farmers or communities which organize and perform as a group in the village with continuous territory.
- ☆ It is applied on-farm activities with crop production, processing and livestock rearing and off-farm processing.
- ☆ In this, each farmer or group of farmers with less than five members are not covered under the system.
- ☆ The system has to opt for third-party certification or needs to join the existing PGS local group.
- ☆ The system is applied in off-farm processing activities like storage, transport and value addition activities by persons/agencies other then PGS farmers.
- ☆ The off-farm input approval in this system is granted by an applicable group to members of the same group which cannot be taken as basis for universal approval for other groups.
- ☆ In this, off-farm inputs require approval by each group for their member's usage.
- ☆ The system ensures traceability only up to end till it is in custody of PGS group.
- ☆ In this, once the product leaves the custody of PGS group there is no control of PGS on its integrity, hence the system is good for local direct sales/direct trade between the producer and consumer and direct trade of packed finished product with system logo between its group and traders/ retailers.

8.5 Documentation and External Audits

8.5.1 Documentation

- ☆ Documentation is the basis for successful management of the farm as operating processes are optimized and legal requirements are met.

- ☆ Descriptive documentation about farmers/certification data is essential for identifying, finding, accessing and using data in future.
- ☆ Data documentation helps in remembering the details of a particular project that may otherwise be forgotten with time.
- ☆ Data documentation is easier and much less time consuming if it is performed at each stage of the process instead of waiting until the end.
- ☆ Farmers use data to make most of their informed decisions, as soil fertility data is used to know more about the type of soil and the need for fertilizer application.

Documentation Required for Certification

1. The certification body should maintain the following documents:
   - ☆ Information related to authority under which Certification Body is conducting its activities
   - ☆ Documented statement of its certification program which should cover policies and procedures for the grant, maintenance, extension, suspension and withdrawal of certification
   - ☆ Information related to inspection, evaluation and certification processes related to each category of certification
   - ☆ Details using which, certifying body gets financial support and information related to fees charged to operators who are desired to be certified
   - ☆ Knowledge related to procedures for handling complaints, appeals and disputes
2. The certification body shall establish and maintain policies and procedures for the creation and control of documents and data which relate to its certification program.
3. The documents should be made available to the evaluation committee who has right to give its feedback and recommendations to certifying body for better document maintenance.
4. The certification bodies shall maintain a system to control all sorts of documentation related to certification system and should make sure that:
   - ☆ The latest issue of relevant documents should be made available.
   - ☆ Correction in documents are made by authorized person
   - ☆ Changes are processed in a manner, which ensures direct and speedy action.
   - ☆ Obsolete documents are removed from further usage.
   - ☆ Certified operators are notified of such changes.

- ☆ Documents shall be reissued on doing substantial amendments.
- ☆ Appropriate documents should be registered and maintained from the respective date of issues.

8.5.2 Plan and Execute External Audits

- ☆ The external audit is basically for food safety which is done by an organization that is not linked with a particular company.
- ☆ External audits can be second party and third party.
- ☆ The second-party audit is done when a customer or supplier audits the company, or when a company audit them.
- ☆ Third-party audit is done by an independent audit organization or government agency.
- ☆ During the external audit, auditors gather objective evidence to show about safety system which:
  - ❖ Should meets the program or customer requirements
  - ❖ Is properly implemented
  - ❖ Works effectively to control food safety hazards

Benefits of External Audits

External audits are a tool that:

- ☆ Provide evidence that the company produces food in a safe environment and is in control of food safety hazards.
- ☆ Promote continual improvement of food safety systems by locating areas of weakness.

Applying for an External Audit

While customers/suppliers initiate second party audit, it could be responsibility to initiate third party audit. Many audit organizations require company to initiate audits, while government agencies often take the initiative. It is recommended that the company should implement their safety system and collect monitoring records for three months before applying for the external audit and should made available following documents:

Application Form

- ☆ Company's Food Safety Management System (FSMS) policy
- ☆ GMP wrote programs, policies, procedures, HACCP plans, blank records.
- ☆ Supporting documents (technical or scientific data to validate critical control points)

8.6 Certification and Traceability

8.6.1 Certification

Organic certification is an implementation system and products control for organic products.

In this, stakeholders who play a role in the process of producing organic products include farmers, processors, and organic products companies who conduct certification to maintain integrity of organic products.

The certification process is conducted by government or private certification organizations that are acknowledged by government, so certification process of organic products comprises of as shown:

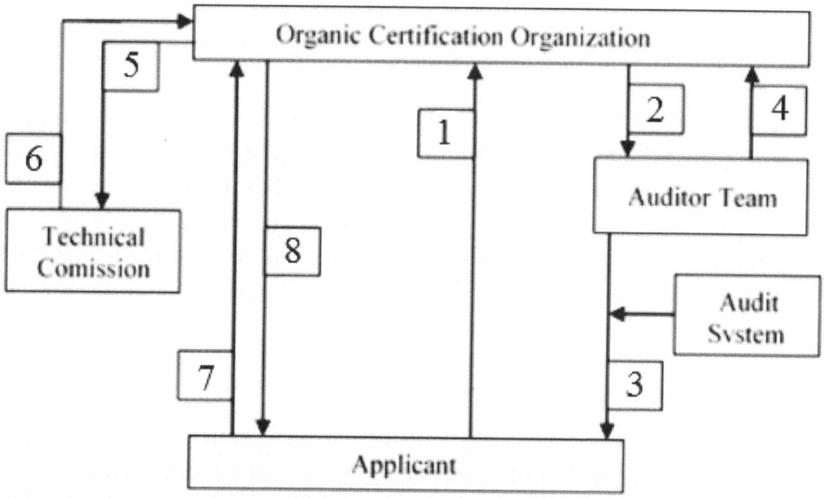

The certification process shows that:

1. Organic farm producers should submit certification proposal to organization of organic product certification.
2. Certification organization appoints an audit team.
3. Audit team conducts auditing process of document adequacy
4. Audit team to deliver audit result to certification organization
5. Certification organization to deliver audit results to certification committee for being discussed in internal meeting for certain recommendations
6. Certification committee delivers recommendations to certification organization.
7. Certification organization to deliver assessment results, whether the applicant deserves certification or not.
8. Organization of organic product certification periodically conducts surveillance.

8.6.2 Traceability

- The demand for traceable and certified agricultural products is on the rise in global markets as international consumers demand more transparency and accountability in supply chains.
- The need to support traceable and certified production is strong in developing countries, where smallholder farmers produce and sell cash crops for international markets.
- While the benefits of traceability and certification are clear for both buyers and sellers of crops in agricultural value chains, there are still significant challenges in implementing such schemes.
- The process of getting certified is long and expensive, for example. Organic certification processes can take up to five years and getting producers to commit the resources to comply with standards can be difficult.
- To comply with an increasing number of standards to meet traceability and certification schemes, companies in the agricultural sector that procure crops from farmers are increasingly looking to implement so-called track and trace systems, a range of mobile-enabled tools that link farmer and farm data to a commodity flow through the related supply chain. These tools enable agribusinesses to have full visibility into the agricultural last mile, maintain a digital record of suppliers and facilitate auditing processes for certification requirements.

Traceability Organic Certification

- In development of traceability information system as certification process, the main thing required is ability of system to save and communicate important documents like cultivation, processing, quality of product, *etc.*
- It is important to define terms of usage requirements for actors which are involved in certification process.
- A system approach was used to develop a certification information system.
- The traceability system was developed to solve the abnormal situation, creating transparency in production process and giving information to consumer and involved stakeholder including food certification institution.

Traceability System in Certification

Record of Cultivation

- Farmers recorded all activities during cultivation.
- The data such as varieties used, irrigation system, plantation, fertilization process, pest and disease of plant *etc.* were recorded.

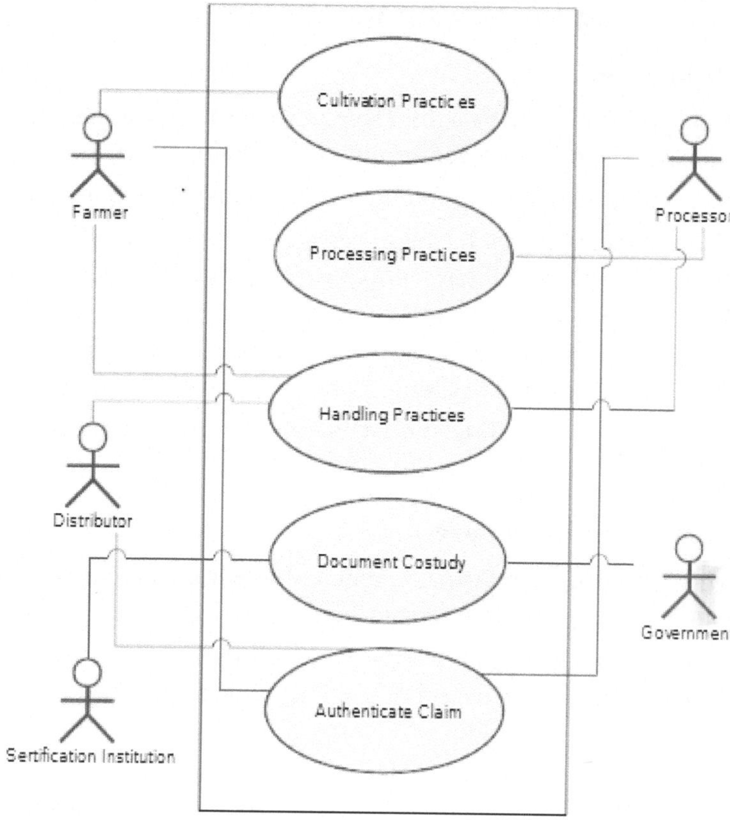

Record of Handling

- ☆ Material handling practices like drying, storage *etc.* conducted by stakeholder, start from farmers, processor and distributor to be well recorded

Record of Processing

- ☆ Processor must record all processing.
- ☆ Many recorded parameters in processing practices includes processed varieties, quantity and quality of product.

Authenticate Claim

- ☆ All actors who play role in organic crop production will take data and claim their activities that conducted has been appropriate with determined organic standard as per storage system.

Document Study

- ☆ Based on demand, certification system should give information holistically about certification actors who has role in organic crop production.

8.7 Procedures and Applicability

8.7.1 Certification

- ✰ It is a procedure by which third party gives written assurance that product, process or service is in conformity with certain standards.
- ✰ It can be seen as form of communication along the supply chain.
- ✰ The certificate shows to the buyer that supplier complies with certain standards, which might be more convincing than if supplier itself provides the assurance.
- ✰ Organic certification addresses growing worldwide demand for organic food and is intended to assure quality and prevent fraud.
- ✰ It is aimed at regulating and facilitating the sale of organic products to consumers.

8.7.2 Certification Procedure

To certify the farm, crop grower is required to engage in many new activities along with normal farming operations such as:

- ✰ Studying about organic standards, which cover in detail what is and is unallowed for every aspect of farming, including storage, transport and sale.
- ✰ Idea about farm facilities and production methods which must comply with standards, involving modifying facilities, sourcing and changing suppliers, *etc.*
- ✰ Involving in a lot of paperwork involving details regarding farm history, current set-up, soil and water tests.
- ✰ Writing annual production plan which needs to be submitted with details of seed and sale: seed sources, field and crop locations, fertilization and pest control activities, harvest methods, storage locations, *etc.*
- ✰ Annual on-farm inspections with the physical tour, examination of records and oral interview.
- ✰ Payment of fee by grower to certification body for annual surveillance and facilitating mark acceptable in the market as symbol of quality.
- ✰ Maintaining records of information related to day-to-day farming and marketing, covering activities for inspection at any time.

8.7.3 Certification Process

The procedure of certification involves following points:

Developing Organic System Plan

- The organic system plan is foundation of organic certification process.
- Crop grower should make the organic system plan by covering details related to operations and regulations as per unique characteristics.
- As plans differ as per operation type and needs, it should address all practices of farming or handling systems like tilling, grazing, harvesting, storing and transporting.
- The plan should cover approved substances used during growing or handling process, monitoring practices for organic systems, recordkeeping systems and barriers which prevents commingling with nonorganic products.

Implementation of Organic System Plan

- The organic system plan needs to be reviewed by certifying agent.
- Organic operations are certified by the private, foreign or State entities that have been accredited by respective council.
- Certifying agents are responsible for ensuring organic products meet all organic standards.

Receiving of Inspection Report

- Every operation that applies for organic certification will be inspected first by certifying the agent.
- Such top-to-bottom inspections differ in scope as per farm or facility.
- For crops, inspection is done for fields, soil conditions, crop health, approach to management of weeds and other crop pests, water systems, storage areas and equipment.
- For livestock, inspection covers feed production and purchase records, feed rations, animal living conditions, preventative health management practices, health records, and number and condition of animals present on the farm.
- At handling or processing facility, an inspector evaluates the receiving, processing, and storage areas used for organic ingredients and finished products.

Review from Certifying the Agent

- The inspector presents findings to certifying agent following observation of practices on the farm or facility as they compare to organic system plan.
- In addition to inspection points, inspector will presents an assessment of risk of contamination from prohibited materials and might even take soil, tissue or product samples as needed.

- ☆ The inspector also analyzes potential hazards and critical control points and makes sure procedures to prevent contamination are adequate.
- ☆ From there all findings are presented the certifying agent for review.

Receiving Decision from Certifier

- ☆ If an operation complies with rules, certifying agent issues an organic certificate which carries the list of products that can be sold as organic from particular operation.
- ☆ The organic farm or facility continues updating its plan as it modifies its practices, and an inspection is done at least once a year to maintain certification.

Chapter 9

Undertake Business of Organic Farming

9.1 Items of Cost in Organic Farming

Cost of Organic Certification

- ☆ It includes major certification costs like renewal certification costs, education, organic land, and livestock from organic origins, organic seed, and special processing equipment.

Cost of Facilities

- ☆ Right from growers to processors, many organic certified operations require special land and/or facilities before the production of food.
- ☆ Organic land costs much more than conventional farmland as there is a list of qualities that organic land possesses.
- ☆ Many organic operations are so small that they don't warrant a full-scope manufacturing facility of their own, either locating an organic operation with which they can share space or purchasing special equipment for a conventional facility.
- ☆ If an organic company shares space with conventional companies, more time is spent in making sure that products aren't mixed, or if they are mixed, then such processing machines are properly cleaned before they're used for organics.

Cost of Small Operations

- ☆ The organic farm costs quite a bit more to grow and process small quantities of organic food, as opposed to large quantities.

- ☆ In addition to added costs to process small amounts of food, it costs more to ship such items.

Cost of Marketing

- ☆ The marketing of business takes a significant amount of money and time.
- ☆ As seen, big companies have set marketing budgets and people who do marketing work for companies, while the small company does marketing operations on their own with a very low budget, which requires considerable creativity.
- ☆ Small companies with or without budgets have to pay for consumer education, print ads, business websites and more, so it involves setting of marketing budget for such companies.

Cost of Synthetics

- ☆ Organic food producers didn't use the same readily available cheap additives, flavors, and preservatives as conventional food producers.
- ☆ As harmful ingredients are not allowed in organic food, so organic producers need to find less harmful others which is also less available and more expensive ingredients.
- ☆ With this, costs will be less to use synthetic food items, so conventional food companies get a real financial break as organic companies don't.

Cost of Labour

- ☆ Conventional growers use all of such chemicals and synthetic pesticides as they wanted to reduce the cost of production by getting the job done faster and more efficiently.
- ☆ Without chemicals, organic growers have to hire more workers for tasks for work related to hand-weeding, cleanup of polluted water, and remediation of pesticide contamination, so cost require is more.

The cost involved in Crop rotation.

- ☆ Instead of using chemical weed-killers, organic growers do crop rotations to keep their soil healthy and to prevent weed growth.
- ☆ Once the organic crop is harvested, organic grower uses that area to grow cover crops, which will add nitrogen to the soil to give benefit to succeeding crops.
- ☆ The conventional crop growers use every acre to grow the most profitable crops as crop rotation reduces the frequency in which organic farmers can grow profitable crops, as they are unable to produce large and cost-effective quantities for conventional farmers.

Cost in Postharvesting Handling

- ☆ To avoid cross-contamination, the organic crop is separated from the conventional crop once it is harvested.
- ☆ Conventional crops are shipped in big quantities as conventional farms are able to produce more.
- ☆ Organic crops are handled and shipped in small quantities as organic farms tend to produce less which results in high costs.
- ☆ The organic farms are located away from main cities, which increases shipping costs.

9.2 Revenue Trends in Organic Farming

- ☆ With the use of digital knowledge in terms of the growth of organic crops, the demand for organic food sees good growth. People are becoming aware of the harmful effects of chemicals and fertilizers used during farming. A great rise in the awareness of organic farming methods makes food more healthy and nutritious.
- ☆ With rising disposable income and access to credit, the middle class will have a high spending capacity on organic food.
- ☆ Due to the following reasons, the demand for organic food products is on the constant increase across the world as the export of organic products was around $575 million which increases in financial years.
- ☆ The major issues in the organic market are mismanagement in the supply chain of organic fruits/vegetables which lead to challenges in terms of cost and logistics by locally or regionally produced organic produce.

9.3 Breakeven Analyses

- ☆ It is a common tool used in evaluating the economic feasibility of a new enterprise or product.
- ☆ The break-even point is point at which revenue is equal to the costs and no profit is made and no losses are incurred.
- ☆ It is a point which can be expressed in terms of unit sales or dollar sales, *i.e.* break-even units show level of sales that are required to cover the costs.
- ☆ Sales above breakeven number will result in a profit, and sales below that number will result in loss.
- ☆ Break-even analysis is based on fixed costs and variable costs where fixed costs are overhead and constant expenses and do not change as the level of output changes while variable expenses are not constant which changes with the level of output as such expenses are stated on per unit basis.

- ☆ After the break-even point is met with no change in selling price, fixed and variable cost, then a profit in the amount of difference in selling price and variable costs can be seen.
- ☆ The equation for determining break-even units is Average Annual Fixed Cost/(Average Per Unit Sales Price - Average Per Unit Variable Cost)
- ☆ The equation for determining break-even sales is Annual Fixed Cost/1 - (Average Per Unit Variable Cost ÷ Average Per Unit Sales Price)
- ☆ Break-even analysis is good for the evaluation of new ventures as mostly new enterprises and products operate at loss in the early stages of development, so knowing about the price or volume is critical to evaluate the time-frame where losses are permissible.

9.4 List of Govt. Schemes

Nowadays, government is giving priority for welfare of organic crop growers by implementing welfare schemes to the re-vitalize organic agriculture sector and in improving economic conditions of growers. The government has rolled out new initiatives, schemes, programmes and plans to benefit organic crop growers. Such schemes or programmes will help crop growers to know about certain benefit, which will be there in respective schemes which include:

Pradhan Mantri Fasal Bima Yojana (PMFBY)

- ☆ It is an actuarial premium based scheme under which growers has to pay maximum premium of 2 per cent for Kharif, 1.5 per cent for Rabi and oilseed crops and 5 per cent for annual commercial/horticultural crops.
- ☆ In this, remaining part of actuarial/bidded premium is shared equally by Centre and State Government.
- ☆ The idea of such scheme is to facilitate the prompt claims settlement where claims need to be settled in two months of harvest subject to timely provision of yield data and sharing of the premium subsidy.

Pradhan Mantri Krishi Sinchai Yojana (PMKSY)

- ☆ The scheme was launched with the idea of providing water to every crop field by the supply chain.
- ☆ The scheme creates protective irrigation by harnessing rain water at the micro-level by 'Jal Sanchay' and 'Jal Sinchan.'
- ☆ The scheme adopts State level planning and productized execution by allowing States to draw up their own irrigation development as per District and State Irrigation Plans.

Paramparagat Krishi Vikas Yojana (PKVY)

- ☆ The scheme has implemented with an idea to promote organic farming in the country.
- ☆ The scheme will help in improving soil health and organic matter content.
- ☆ The scheme will increase the net income of crop growers so as to realise premium prices.

National Agriculture Market (e-NAM)

- ☆ The scheme provides e-marketing platform at national level.
- ☆ It supports the creation of infrastructure to allow e-marketing.
- ☆ Such type of innovative process will revolutionize agriculture markets by allowing better price discovery.
- ☆ The scheme brings transparency and competition to allow crop growers to get improved remuneration for their produce./

Rainfed Area Development Programme (RADP)

- ☆ The scheme will improve quality of life of crop growers by offering complete package of activities to maximize farm returns.
- ☆ The scheme increases agricultural productivity of rainfed areas in sustainable manner by adopting appropriate farming system based approaches.
- ☆ It will minimise adverse impact of possible crop failure due to drought, flood or un-even rainfall distribution by diversified and composite farming system.
- ☆ The scheme will rectorate confidence in rainfed agriculture by creating sustained employment opportunities by improved on-farm technologies and cultivation practices.
- ☆ The scheme will upgrade crop growers income and livelihood support for reduction of poverty in rainfed areas.

National Mission for Sustainable Agriculture (NMSA)

- ☆ The scheme will promote Sustainable Agriculture by climate change adaptation measures and improves agriculture productivity in rainfed areas with focus on integrated farming, soil health management and synergizing resource conservation.
- ☆ The scheme focuses on conservation agriculture to make farm sector more productive, sustainable, remunerative and climate resilient by promoting location specific integrated/composite farming systems.

Gramin Bhandaran Yojna

- The scheme creates scientific storage capacity with allied facilities in rural areas.
- The scheme was launched to meet the needs of crop growers for storing farm produce, processed farm produce and agricultural inputs.
- The scheme will help in promotion of grading, standardization and quality control of agricultural produce to improve their marketability.
- The scheme will help in preventing distress sale soon after harvest by using facility of pledge financing and marketing credit by strengthening agricultural marketing infrastructure.

Mission Organic Value Chain

- The scheme will create at least one to two replicable end-to-end organic value chain models in state with integration of growers, handlers, processors and market facilitation agencies.
- The scheme will give an idea of converting of subsistence farming to commercial organic farming with end-to-end facilities.
- The scheme will help in improving the production system to allow high productivity with better profitability.

9.5 Market Trends and Consumers Choices

9.5.1 Trend in Market

- The Organic Farming industry has benefited from a shift in consumer preferences towards organic produce over the past five years.
- Demand for organic products has risen over the period as consumers have increasingly considered the health and environmental consequences of their food choices.
- Organic farming is a method of crop and livestock production that is free of synthetic chemicals, pesticides, fertilizers or genetically modified organisms.
- Organic farms produce organic fruits, vegetables, grains other crops or livestock.
- The Organic Farming industry consists of small operators, which makes it difficult to maintain consistency in the quantity and quality of produce.
- Presumably, organic food contains fewer chemical residues and veterinary drugs as compared to conventional food with only one-third of pesticides.

9.5.2 Consumer View

- Many consumers assume that they cannot afford organic food while some feel that the market often charges more for healthier food.
- The consumers want to have a choice among sustainable products rather than choosing sustainable products and those that are not.
- Consumers information about product attributes and consequences to evaluate a product and make their choices as they rely on their felt involvement which is influenced by experience.
- The importance placed on each parameter is based on consumers' priorities and values.
- Consumers are becoming more conscious about what they buy as they're choosing an organic product as they have more knowledge of how it guarantees higher welfare standards, contains fewer pesticides with no added artificial additives or chemicals, and is better for the planet which meets the demand for less packaging and products from sustainable farming sources.
- Increased consumer demand, supply chain improvements and improving economies of scale are all anticipated to drive industry revenue growth over the next couple of years.
- Organic consumers are willing to pay around 10 per cent premium for organic food with an average of 9.5 per cent by women and 11.4 per cent by men.
- Regular consumers pay a slightly higher premium around 15 per cent, where average of 12/6 per cent premium is paid by women while 18 per cent premium by men.
- Organic food has been attaining growing consumer demand due to health factors followed by environmental and animal welfare factors, so consumer perceptions about organic food are highly subjective.
- It is noting that such perceptions will not be their actual behavior in purchasing as good quality of organic food at a reasonable prices not only attract more buyers but also do justice towards the environment due to less usage of pesticide applications.

9.6 Disseminating Benefits of Organic Food

- Organic farming has attracted attention as consumers increasingly prefer the putative safety and health benefits of organic products.
- Cultivation of productive crops by Green Revolution has necessitated the usage of many fossil energy resources for use of agricultural machinery, chemical fertilizers, synthetic agricultural chemicals, and other inputs

which led to environmental problems like reduced biodiversity from chemicals, soil and water pollution.

- crop growers are deeply interested in organic farming to get product unit prices which are more than products grown by using chemical fertilizers, which lead to low yields of organic farming.
- The standards on Organic Food Systems states that the use of chemical fertilizer and pesticides to be avoided, which disqualifies paddy fields practice organic farming in rainy season while conventional cultivation by chemical fertilizers in the dry season for organic farming certification.
- As producers are increasingly reducing the environmental burden and maintaining food safety, the development of human resources and agricultural extension centers will be required to protect sincere producers.
- Composting waste materials and using them for organic farming contributes to measures against environmentally conscious farming practices and leads to the development of sustainable recycling agriculture.

9.7 Organic Produce Market Demand and Prices

9.7.1 Introduction

- Consumers prefer organically produced food due to their concerns related to health, environment, animal welfare which shows a willingness to pay price premiums as established in the market place.
- Organic products shift from being a lifestyle choice for a small share of consumers to be consumed occasionally by the majority of people.
- Rise in lifestyle diseases like diabetes, blood pressure, obesity and excessive contamination of conventional food drives organic food sales, worldwide.
- Due to the regular increase in demand for organic food and limited supply in relation to demand of organic foods, the area under organic farming has increased worldwide.

9.7.2 Organic Food Market

- Consumer demand for organically produced goods continues to show double-digit growth, providing market incentives for crop growers across a broad range of products.
- Organic products are available in natural food stores and conventional grocery stores.
- Organic fruits and vegetables are the largest and fastest-growing segment of the global organic food market.
- The upsurging demand for organic fruits and vegetables is mainly due to its increased production and growing consumption across the globe.

- ☆ The organic fruits and vegetable supply is limited in relation to its demand, due to its high production cost.
- ☆ In spite of the high price for organic fruits and vegetables, consumers are willing to pay more for organic foods due to less usage of pesticides and high nutritive value.
- ☆ Increasing awareness about health benefits from organic products among the consumers boosts the organic fruit and vegetable sales in the market.
- ☆ The global organic food market is at around $110.25 billion which grows at 16.15 per cent CAGR which increases to $262.85 billion.

9.7.3 Rising Demand for Organic Food over Conventional Food

- ☆ Rapid urbanization, increased adoption of healthy dietary lifestyle and rising concerns regarding health and environment are the main factors that attributes to the growth of organic fruits and vegetable markets.
- ☆ The rise in lifestyle diseases like diabetes, blood pressure and obesity with excessive contamination of conventional food, drives the organic food sales across the globe.
- ☆ The constantly rising demand serves as an opportunity for supermarkets and food store chains for diversification and widening of offerings of organic food at competitive prices.

9.7.4 High Production Costs and Low Yield

- ☆ The production costs of organic fruits/vegetables are higher than conventional fruits and vegetables.
- ☆ Organic food costs around 20 per cent - 100 per cent more than conventional produce.
- ☆ Organic foods are difficult to grow due to high involvement and growing time, but low yields of organic crops and poor supply chain led to rise the production cost.
- ☆ The major hurdles in the adoption of organic farming are a high fee for registration and getting accreditation and certification to become organic farmers.
- ☆ Owing to low organic production yield as compared to conventional yields, there will be an increase in demand for extra-natural resources such as land, water in order to meet, the growing demands of organic fruits/vegetables.
- ☆ Organic fruits/vegetables are the largest and fastest-growing segment of the global organic food market.

9.7.5 Indian Organic Food Market

- ☆ Indian organic food market grows at about CAGR of above 23 per cent by 2023 due to favorable conditions and cultivation.
- ☆ Agricultural and Processed Food Products Export Development Authority (APEDA) shows that the export of organic products is around Rs.30 billion ($440 million).
- ☆ With more awareness and a rise in demand for organic food, the sales increase.
- ☆ As buyers are more aware of the harmful effects of chemicals and pesticides, so they have started looking for organic products.
- ☆ With an increase in disposable income and awareness, consumers of organic products are spending more and more money on their well-being.
- ☆ The consumers are ready to pay higher/premium prices of organic products for the sake of product quality.
- ☆ Agricultural and Processed Food Products Export Development Authority highlights that the demand for Indian organic food products is on the constant increase across the globe which priced around $515 million in the last financial year.

9.7.6 Pricing

- ☆ Fresh fruits/vegetables have been top selling category of organically grown food as organic food industry started retailing products over 3 decades and are outselling other food categories.
- ☆ The challenges that come in the selling of organic fruits/vegetable is mismanagement of the supply chain as it becomes difficult to fulfill demand as per requirement.
- ☆ There are many costs linked with the procurement of organic food from organic farms of states in terms of logistics, damages and remaining stock that doesn't receive a premium price.
- ☆ High price markups for organic products as compared to conventional products are the main factors that affect the sale of organic products.
- ☆ While the high price is due to cost linked with products such as procuring in bulk, so the logistics cost involved in the procurement of organic products from certified organic farms and distribution in the city lead to increase the product cost.

9.8 Major Branding Methods of Organic Produce

9.8.1 Branding

- ☆ The brand is a mark or identifier, on a product or service, which allows it to be recognized by individuals.

- ☆ The value of branding to originators like multinational organizations or individuals emerges from diverse areas.
- ☆ Branding provides value to consumers/companies who are often potential buyers of the branded products.
- ☆ The potential value of produce emerges from brands that create a sense of being different from alternatives in meaningful ways and provides assurance or consistency of quality.
- ☆ Brands that are successful, have the potential to reduce the effort from the audience which is required to search for and evaluate alternative brands.
- ☆ Prominent and enduring brands incorporate mutually supportive features, which helps in creating a brand image that forms a unique position from the audience's perspective, relative to alternative offerings.

9.8.2 Branding of Organic Products

- ☆ Organic products gained much popularity in recent years and people came to know that artificial items are not good for health and nature in the long-term.
- ☆ Organic products are made from natural ingredients only and are fresh to consume which makes them beneficial for us.
- ☆ Perceived risks pertaining to food consumption and lack of knowledge in terms of organic products are leading consumers to rely on indicators like brand name, store image, label or partners like producers.
- ☆ A product differentiated with intangible feature as organic is a credence good which pose challenges for consumers, overcome with the creation of the brand.
- ☆ Credence attributes are difficult for consumers to judge even after purchase, in contrast to search attributes; which can be evaluated before purchase, and experience attributes that can be evaluated after purchase, but it is challenging to create one brand for all organic products, as there are many independent organizations who collectively make up organic food movement.
- ☆ A product differentiated with intangible feature as organic is a credence good which pose challenges for consumers, overcome with the creation of the brand.
- ☆ The potential power of the organic brands has been diminished by the usage of 'organic' on products that are not certified organic.

9.8.3 Brand Building

- ☆ A successful brand has to be consistent in communication and experience, across many applications such as:

- ❖ Environment
- ❖ Print, signage, packaging
- ❖ Website and online advertising
- ❖ Content publishing
- ❖ Sales and customer service

- ☆ Building a brand is a process, but ongoing effort results in establishing long-term relationships with customers which leads to a steady increase in sales, more projects, word-of-mouth referrals and advocacy for products or services.
- ☆ Brand building generates awareness about the business using marketing strategies and campaigns with the aim of creating a unique and lasting image in the marketplace.

Building an organic produce brand involves:

Determining of Brand's Target Audience

- ☆ The foundation for building the brand is to determine the target audience that you'll be focusing on.
- ☆ While building a brand, crop produce manufacturer/owner should keep in mind, who exactly needs to be targeted.
- ☆ Brand creation relies on an understanding of buyer persona such as age, gender, location, income, level.
- ☆ Identifying target customers for the product is an exercise that will affect and benefit all areas of the brand-building process, particularly marketing efforts.

Establishing a Brand Mission Statement

- ☆ The mission statement defines the purpose for existing which informs every other aspect of brand building strategies.
- ☆ It covers everything from the logo to tagline, voice, message, and personality which should reflect particular mission.

Researching of Brands within the Industry

- ☆ Crop produce manufacturer/owner should do research on main competitors or benchmark brands in order to build the brand name.
- ☆ For brand name to be effective, it needs to be easy for consumers to recognize and remember.

Creating Brand Competitor Research Sheet

- ☆ Competitor research is the main element of making brand development.

- ☆ Crop manufacturer/owner should start by making brand competitor sheet in order to compare.
- ☆ The crop produce manufacturer/owner should use Google sheets, Excel or simple notebook to write details regarding answers to questions as:
- ☆ Is the competitor consistent with messaging and visual identity across channels?
- ☆ What is the quality of competitor's products?
- ☆ Does the competitor have customer reviews or social mentions?
- ☆ In what ways does the competitor market their business, both online and offline?

Outline Main Qualities and Benefits of the Brand Offered

- ☆ Assuming that crop produce manufacturer/owner know exactly who their target audience is, give them a reason to choose the brand over another.

Creating Brand Logo and Tagline

- ☆ The exciting of brand building process involves creation of the brand logo and tagline for the company.
- ☆ The logo should appear on everything that relates to business as it will become an identity, calling card, and visual recognition of promise.

Building Brand Voice

- ☆ The voice is dependent on company mission, audience and industry as it shows how one will communicate with its customers and what will be the response.
- ☆ Brand voice is mainly important while publishing blog articles or social media posts.
- ☆ Maintaining consistent voice will help in brand image to become recognized on multiple channels in a similar way.

Building Brand Message and Elevator Pitch

- ☆ When building a brand, tell customers succinctly who you are.
- ☆ Building brand message and elevator pitch defines key aspects of who you are, what you offer, and why people should care.
- ☆ Brand message is an opportunity to communicate on human level which makes direct emotional connection with the consumers.
- ☆ While creating a brand message, crop produce manufacturer/owner should address not what the product can do, but should stress on, why the product is important to the customer.

9.9 Major Retailers, Wholesalers and Online Marketplaces for Organic Produce

9.9.1 Marketing of Organic Produce

- Marketing is an important factor in determining the success of the organic crop or related product.
- It includes all operations and decisions made by producer/grower who can be determination of most marketable crops for production, best delivery quality produces, *etc.*
- Marketing does not begin after crop is produced, but is an alternatives need to be considered even before production takes place.
- Many organic businesses sell to one buyer while majority sell to number of buyers, into quite different markets.
- When crop is supplied to the fixed buyer every year, then crop grower becomes part of integrated supply chain.
- Organic crop growers sell part of their produce on the spot market, where the last one is more for speculators.

9.9.2 Marketing System

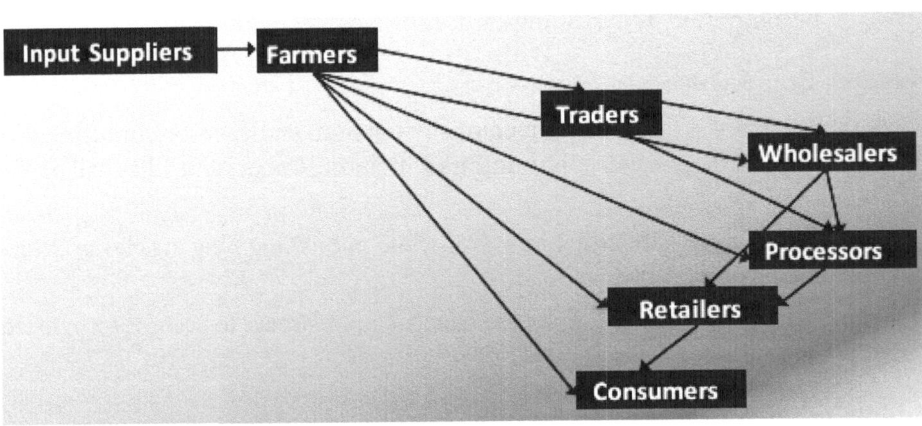

- In India, the current marketing system for organic crop includes private and public system, traders and processors.
- There are number of stakeholders involved in agricultural commodity value chains such as farmers, traders, wholesalers, retailers, big retail chains and consumers are major actors of the vale chain.

Traditional Organic Markets

- Local markets

- Farmers markets
- Local shops, restaurants, hotels *etc.*
- Farm Gate Sales
- Home delivery
- CSA
- Food Coops
- Agents and Merchants at Wholesale Markets

The most common marketing channels includes;

Private

The Marketing channels in private sector are –

- Farmer– Consumer
- Farmer– Retailer or village trader– Consumer
- Farmer– Wholesaler– Retailer– Consumer
- Farmer– Village Trader– Wholesaler– Retailer– Consumer
- Farmer– Wholesaler– Miller– Retailer– Consumer

Institutional

- It covers the public and co-operative sector agencies.
- It plays a very important role in procurement and distribution of organic crops.
- The Food Corporation of India is main agency for procurement, maintenance of buffer stocks and distribution of organic crops.

The main institutional marketing channels for organic crops includes;

- Farmer
- Shop– Consumer
- Farmer– Co-operative Marketing Society– Retailer– Consumer
- Farmer– (FCI, State Govt., Co-operatives)– Private Trader– Export
- The system includes the Commission on Agricultural Costs and Prices (CACP), Food Corporation of India (FCI), State Civil Supplies Corporations/Departments.
- The CACP studies costs, prices, and markets recommend prices to the national government.
- The national government's grain management includes procurement, transport, storage and release of grain mainly by FCI.

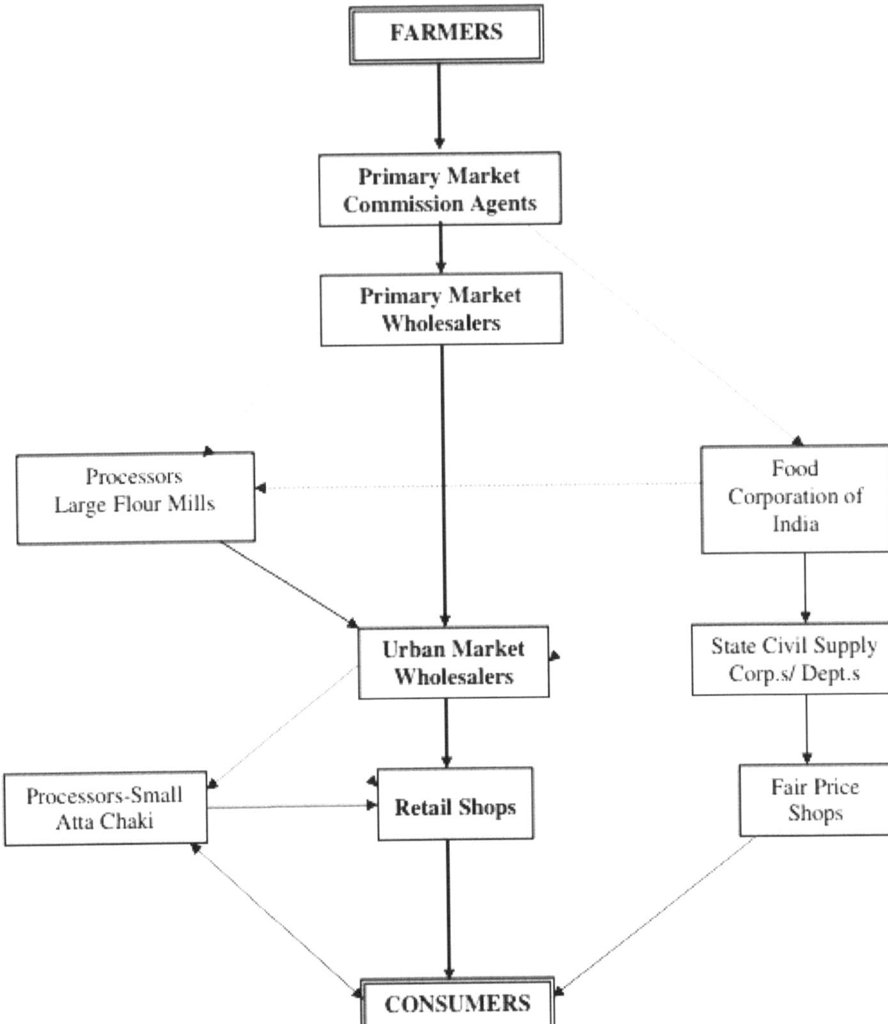

Marketing System in India.

- ☆ Grain distribution is done by the public distribution system through state governments and fair price shops (ration shops) that spread across the country in both rural and urban areas.
- ☆ Over many years, more active minimum support price (MSP) and procurement policies of the Government have reduced the role and activities of private traders.
- ☆ There is very little private stocking of wheat as the government purchases and stocks a large amount of wheat.

- The uncertainty regarding the price and volume of sales from large government stocks often makes it difficult for private traders to play their role in the market in a significant way.
- Many consider the current situation a de facto nationalization of wheat trade.
- The government's purchase and sale policies, together with stock holding and movement restrictions under the Essential Commodities Act, state policies, taxation policies, have inhibited private investment in improving grain marketing infrastructure.
- The inefficiency of Food Corporation of India (FCI) exits in terms of handling, transporting and storing of government grain.
- As observed, the government is more efficient than private traders, while with FCI setting the base, there might be less incentive for private traders to be more efficient.
- As there appears the main role of wheat/rice in farm income and consumer expenditures, inefficiencies in marketing would impose significant costs on producers, consumers, and the economy.

9.10 Individual vs the Group Marketing

9.10.1 Marketing Information

Marketing of agriculture produce includes assembling, handling, storage, transport, processing, wholesaling, retailing and export of agricultural commodities with supporting services like market information, the establishment of grades and standards, commodity trade, financing and price risk management.

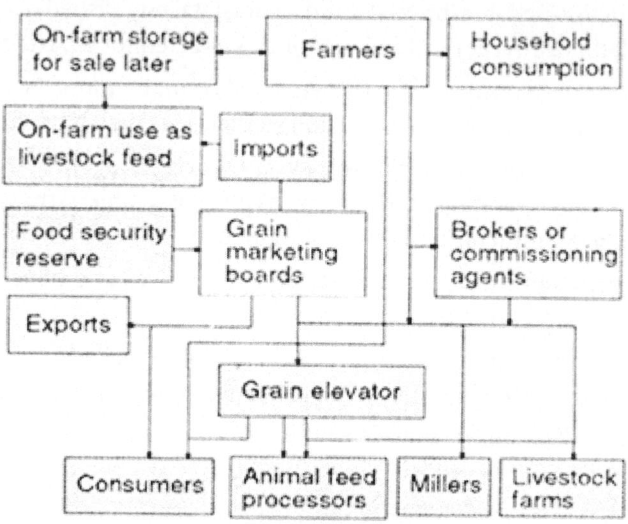

Marketing System.

- The marketing system starts with assembling and collecting points located in rural areas close to the growers/producers.
- The next stage involves storage areas at national grains marketing facilities owned and operated by parastatal and/or private grain elevator, and grain milling companies which in many countries are privately owned and in others are government enterprises.
- Assize and methods of operation differ from country to country, local assembling and collection points will have grains directly by the farmer-producers themselves or by rural entrepreneurs.
- In grains, assembly and storage functions are typically combined at the marketing stage.
- In countries where the marketing parastatal is in monopoly in grain trading, private traders are sometimes authorized to buy grains from farmers on behalf of the parastatal.
- A common feature of grain marketing systems is the co-existence of government marketing agency (parastatal) and parallel private marketing channels with myriads of private traders.
- Public grain marketing agencies are government-appointed parastatals assigned to control or regulate the system.
- Earlier than market liberalization, marketing parastatals were recognized in many developing countries that are responsible for setting the prices for major cereals.
- As per the country, agencies normally consist of one or more ministries related to cereals production and influencing public policies that affect food production and consumption and a parastatal established to operationalize regulatory provisions of public policies enacted for a particular crop.
- In many developing countries parastatals remain important players in the grain marketing system even though their role may have changed.
- In the commodity marketing system, private agents include private individuals operating in the system as petty assemblers, traders large-scale merchants, millers, brokers, and retailers of grain products.
- The agents in the channel operate in rural areas and penetrate remotest areas of rural areas to purchase grains.
- In developing countries where national structural adjustment programs have not yet been initiated, inter-provincial grain movement controls show private agents in the system cannot legally and profitably transport grain from one production zone to another.

- For countries that are adjusting their economies structurally, market liberalization is making or will soon make many restrictions on commodity movement irrational, and new forms of intermediary commodity movement will come into being.

9.10.2 Group Marketing

- It is a method used by crop growers and farmers to market different organic crops.
- It aims to deliver messages to a large number of people.
- It ignores the market segment differences and provides one offer to the whole market.
- In this, farmers/crop growers form a group for the marketing of their crops/produce
- In this, farmers/crop growers take advantage of common transportation, storage, access to information, bargaining
- The groups can approach for backward linkage and forward linkage.
- The groups can directly sell their produce to hotels, hostels, restaurants, *etc.*
- The groups can take advantage of certain Government schemes.
- Reaching a wide audience and maximizes the product's exposure.
- The marketing focuses on high sales and low prices.
- It's used by companies that offer products that usually aren't dedicated to one client and are sold in large quantities.
- Due to its broad audience and low production costs, it is a good way of marketing.
- It has high advertising costs and high development costs of products.

9.10.3 Individual Marketing

- It analyzes the specific needs of each individual and reaches the target group by providing tailored offers of products and services.
- It comprises customers identification, segmenting clients into more specific groups, choosing the best communication channel to contact them and personalizing products or services for giving exactly what they want.
- Such marketing is distinguished by certain advantages like particularly targeted customers which tends to be loyal, less competitive and higher profit margins.
- The drawbacks of such marketing could be a lack of economies of scale and higher production costs per each person.

9.11 Long Term Benefits of Organic Farming Business

- Organic farming opens a new niche market for crop production.
- As consumer demand is growing for organic produce, the relative price of organic produces increases day by day and once the grower successfully transitioned the farm to be organic, it takes only 3-5 years to drive up the yields to their previous rates.
- Organic farming provides substantial benefits to farmers and consumers across the nation and can be profitable which appeals to consumers as a healthy and ethical option.
- Organic farming reduces production cost by 25-30 per cent as it does not involve in the use of synthetic fertilizers and pesticides which makes organic farming cost-effective, hence they result in definite benefits to farming with a long term which covers features such as:
    - It is a method that uses the natural environment to upgrade the productivity of agriculture.
    - It uses carbon-based fertilizers and biological pest control and does not use synthetic fertilizers.
    - It carries crop diversity as in conventional farming, mass production of one crop in a location is focused while in organic farming it has multiple crops in the same place.
    - It uses a method such as green manure and composting which replaces nutrients taken from soil from previous crops and relies on the natural breakdown of organic matter which allows the production of nutrients in the soil.
    - It improves soil fertility and feeds nutrients to the soil to feed the plant.
    - It controls the organisms using methods like biological pest control and Integrated Pest Management.
    - It improves the health of people, soil and ecosystem.
    - It reduces the use of nonrenewable energy sources by using organic fertilizes thereby reducing the consumption of agrochemicals.
    - It helps in controlling the greenhouse effect.
    - It stops global warming and climate change and keeps the environment clean.
    - With the use of fertilizers and pesticides, groundwater gets polluted, but with the use of organic fertilizers, water pollution gets prevented.

- With organic farming ecological services like soil forming and conditioning, waste recycling, nutrient cycling, stabilization of soil, carbon sequestration, predation, pollination, and habitats can be improved.
- It improves the taste and quality of food as organic agriculture are climate and ecologically friendly.

Chapter 10

Maintain Health and Safety at the Workplace

10.1 Undertake Basic Safety Checks before Operation of all Machinery and Vehicles

10.1.1 Injuries Due to Farm Machinery/Vehicle

Farm equipment and vehicles daily check is simple and effective ways to spot potentially dangerous issues or defects before they are used. Crop growers or operators should need to have a system of daily routine checks for machinery and types of equipment that are used in farms to ensure that it is in good working order, safe and fit for purpose before the start of working day or work shift.

Crop growers and farm workers use tractors and other equipment every day. These machines are powerful, and if not operated safely, can lead to disabling injuries or death. Modern tractors, telehandlers, and other farm equipment have given farmers the ability to work more quickly and efficiently, but this machinery brings new dangers.

These agricultural machines have potentially dangerous moving parts, which can cause serious or fatal injuries such as:

- ☆ Balers – pick-ups, twine mechanisms and moving rear doors;
- ☆ Forage harvesters – chopping cylinders;
- ☆ Combine harvesters – augers in the grain tank and the header unit;
- ☆ Potato harvesters – rotating rollers and conveyors;
- ☆ Slurry tankers – power take-off (PTO) shafts;

- Bale and straw choppers – chopping mechanisms;
- Tractor – hitch mechanisms, PTOs and PTO shafts;
- Power harrows – rotating tines;
- Feeder wagons – rotating components in the mixing chamber.

Poorly used or faulty machinery is the main cause of death and injury on farms. Machinery accidents can be prevented by keeping the machine in good repair, fitting and ensuring all safety equipment like guards, safe access platforms and ROPS to be operational with the machine all the time.

Crop growers and farm workers before using the machines or vehicles should ask themselves:

- Is the machine they intend to use is suitable for the job?
- Are all safety devices like guards are in place and are working correctly?
- Are machine operators properly trained to do this job/use this machine?
- Has the instruction manual for the machine been provided, read and understood?
- Is the right personal protective equipment is available and worn?
- Has a risk assessment been carried out?
- Has the work been properly planned and communicated to those who may be at risk?
- If two or more people are involved, has everyone understood what needs to be done and has a system of communication been agreed on?
- Are all operators/workers trained to do this job?
- Is suitable clothing and footwear available and worn (including any PPE)?
- Have operators been provided with relevant information?

10.1.2 Checking the Machine before Use

Before working with any machinery, crop growers/operator should carry out the basic check to make sure that the machine/vehicle what he/she will be using is in the good working order and is safe for use. For this, the machine/vehicle should be checked with basic checking like:

- Mechanical defects – pay particular attention to items like brakes, wheels, and tires.
- Guards and other protective devices should be correctly fitted and maintained in good condition.
- Check that Guards and other protective devices are incorrect place, securely attached and working properly.

- ☆ The damaged or defective guards/devices should be repaired or replaced before the machine is used.
- ☆ Check that stopping devices are functioning correctly, such as emergency stops.
- ☆ See that all controls should be clearly marked to show what they do.
- ☆ If workers are to be carried on the machine, ensure that they can do so safely, eg check for safe means of access and that working platforms have guard rails, *etc.*
- ☆ Hitching and attachment points – check that mounted, semi-mounted or trailed machinery are safely attached to the towing vehicle such as a tractor.
- ☆ Examine the conditions of drawbar/pick-up hitch, hitch rings, pins, clips, *etc.*
- ☆ Carry out any pre-use checks as specified in the operator's manual.
- ☆ For self-propelled machines, make sure that mirrors are clean and properly adjusted.
- ☆ Check any other reversing aids are working.

Checking Grain Harvesting Equipment: Safety before Starting

Crop growers/operator should check for following before using grain harvesting equipment:

- ☆ Before trying to operate a combine, the operator/farmer should study the operator's manual which carries information about all general safety rules with specific safety recommendations for the particular machine.
- ☆ Farmer/operator if checking the combine inside the building should make sure that all doors are opened to provide good ventilation as exhaust fumes from gasoline or diesel engine are poisonous which can lead to a serious health hazard.
- ☆ Always clean the combine before starting as trash around the exhaust system can cause fires.
- ☆ Clean the oil, grease or mud on ladders or on the platform of the combine as they can cause serious falls.
- ☆ Check if the combine is equipped with a cab, then clean the glass to provide maximum visibility.
- ☆ Check the tire pressure of the combine every day as under-inflation can lead to buckling of the sidewall, which causes dangerous tire failure.
- ☆ Check the brakes once in a week as with hydraulic brakes, it is required that the master cylinder should have full fluid and should be free from the air inside the lines.

- ☆ Check the threshing cylinder-rocking bar to see it is clear of the cylinder.
- ☆ Make sure that all shields and covers are in place and fastened securely.
- ☆ Remove or stow all service equipment.
- ☆ Always use the handrails and ladders provided on the combine for the safe mounting and dismounting.
- ☆ Be sure that all PTO covers, safety stands, and shields are on the machine before taking it to the field.

Checking Bailing Hay Equipment

Crop growers/operator should check for following before using the bailing hay equipment:

- ☆ Check that the baler should be properly connected to the tractor.
- ☆ Bailer should have required safety guards fitted.
- ☆ See that nobody should be allowed to ride on the baler.
- ☆ See that nobody is getting too close to the baler.
- ☆ Check that the fire extinguisher should be fitted to the machine.
- ☆ Disengage PTO and apply flywheel brake before doing any baler adjustments.
- ☆ Stop the engine and apply flywheel brake before repairs or stringing up the baler.

10.2 Reporting Hazards to the Appropriate Supervisor

10.2.1 Risk at Farms

Risk is a chance that someone could be harmed by hazards like working from ladders or working around electricity. A risk assessment is an important step in protecting people, business and complying with applicable laws.

Crop growers and agriculture workers are particularly exposed to ergonomic hazards such as:

- ☆ Tiring or painful positions
- ☆ Carrying or moving heavy loads
- ☆ Standing or walking and repetitive hand or arm movements
- ☆ Work-related musculoskeletal problems

Farmers live with risk and make decisions every day that affect their farming operations. Many factors that affect the decisions which farmers make could not be predicted accurately such as:

- ☆ Change in weather conditions

- Drop in prices at the time of harvest
- Non-availability of hired labor during working times
- Breakage of machinery and equipment when required
- Dying of draught animals
- Change in government policies

The above changes are examples of risks that farmers face in managing their farm business. Such risks affect farm profitability.

10.2.2 Sources of Risk

The most common sources of risk in farming can be in areas of:

- Production: Production risk occurs from the uncertainty of factors that affects the quantity and quality of farm produce.
- Marketing: Marketing risk occurs due to the variability of product prices and uncertainty of future market prices.
- Financial: Financial risk occurs when money is borrowed to finance the farm business.
- Institutional: Institutional risk occurs due to unpredictable changes in the provision of services.
- Human: Human risk involves risk to farm business that is caused by human illness and the personal situation of farmer's family.

Based on crop grower's attitude, risk divides farmers among:

Risk-Averse Farmers

- Risk-averse farmers avoid taking risks.
- They tend to be more cautious individuals with preferences for less risky sources of income.
- They sacrifice a certain amount of income to reduce the chance of low income and losses.
- They do not refuse to accept any risk at all.
- The risk-averse farmer seeks to be compensated for the risk taken by receiving a higher return than would normally be obtained if there were no risks.

Risk-Taker Farmers

- Risk-taker farmers are people who are open to more risky business options.
- They choose the alternatives which give some chance of a higher outcome, even though they may have to accept a lower outcome.
- They are influenced by the return they could receive.

Risk-Beutral Farmers

- Risk-neutral lies between the risk-averse and risk-taking positions.

10.2.3 Work of Site Supervisors

Supervisors are responsible for the day to day incidents or hazards that could arise at the workplace. He/She must ensure a safe and healthful workplace for crop growers/workers. The following are the main responsibilities that supervisors have in the area of safety and health for workers under their supervision which includes:

Conducting Orientation and Training

- Supervisors should train and instruct their workers, so they can perform their work safely.
- He/she should make the workers understand the type of personal protective equipment required for every work and should train the workers about the proper usage, storage, and maintenance of the equipment.
- In case of safety training courses, the supervisor should ensure that the workers should take the particular course.

Enforce Safe Work Practices

- Supervisors responsibility is to enforce the safe work practices and procedures, the failure of which, will lead to accidents.
- They should encourage workers to identify unsafe or unhealthful workplace conditions or hazards.

Correct Unsafe Conditions

- Supervisors' must take immediate steps to correct unsafe or unhealthful workplace conditions or hazards within their authority and ability to do so.
- In case of unsafe or unhealthful workplace condition or hazard which cant be immediately corrected, in such case, the supervisor should take temporary precautionary measures.
- He/she must follow-up to ensure that corrective measures are completed in a timely manner to address the hazard.

Prevent Lingering Unsafe or Unhealthful Workplace Conditions or Hazards

- There are many incidents which are caused by unsafe or unhealthful workplace conditions or hazards.
- It's the supervisor's responsibility to train and periodically remind workers of what to look for and how to correct or report unsafe conditions or hazards.

- If a hazard is identified, the supervisor must act on it.

Investigate Workplace Accidents

- Supervisors are responsible for conducting accident investigations and for ensuring that all occupationally injured employees report to the Occupational Medical Service immediately.
- He/she should note all facts and opinions regarding the cause of accident.
- Supervisors must review the circumstances, sign and submit the forms within 48 hours.

Promote Quick Return to Work

- Employees must be encouraged to return to work as soon as possible.
- The longer, an employee, is away from work, the less likely he or she will actually return.
- When possible, light or limited duties should be identified and considered, to assist in returning the employee to work.

10.2.4 Reporting Hazards to Supervisor

- Crop grower should be able to report unsafe or unhealthful workplace conditions or hazards to their supervisor without any fear.
- Hazards can be reported verbally or by filling a simple form which can be made available from the store, bulletin boards or other conspicuous places.

Hazard Report Form - Example	
Name:	Date:
Location:	
Equipment:	
Description of the hazard:	
Suggested corrective action:	
Signature:	
Supervisor's remarks:	
Corrective action taken:	
Signature of Supervisor:	Date:

- The hazard reporting procedure allows for prompt reporting and subsequent corrective action without waiting for the next round of regular inspections.

10.3 Use of Farm Machinery and Equipments

10.3.1 Understanding/Checking Machine before Use

Before working with any machinery, the crop grower should carry out the basic check to make sure that the machine/vehicle that he/she will be using is in good working order and is safe for use. For this, the machine/vehicle should be checked with basic checking like:

- Mechanical defects – pay particular attention to items like brakes, wheels, and tires.
- Guards and other protective devices should be correctly fitted and maintained in good condition.
- Check that Guards and other protective devices are incorrect place, securely attached and working properly.
- The damaged or defective guards/devices should be repaired or replaced before the machine is used.
- Check that stopping devices are functioning correctly, such as emergency stops.
- See that all controls should be clearly marked to show what they do.
- If workers are to be carried on the machine ensure that they can do so safely, *e.g.* check for safe means of access and that working platforms have guard rails, *etc.*
- Hitching and attachment points – check that mounted, semi-mounted or trailed machinery is safely attached to the towing vehicle such as a tractor.
- Examine the conditions of drawbar/pick-up hitch, hitch rings, pins, clips, *etc.*
- Carry out any pre-use checks as specified in the operator's manual.
- For self-propelled machines, make sure that mirrors are clean and properly adjusted.
- Check any other reversing aids are working.

Checking Harvesting Equipment

Crop grower should check for following before using grain harvesting equipment:

- Before trying to operate a combine, the operator/farmer should study the operator's manual which carries information about all general safety rules with specific safety recommendations for particular machine.

- ☆ Farmer/operator if checking the combine inside the building should make sure that all doors are opened to provide good ventilation as exhaust fumes from gasoline or diesel engine are poisonous which can lead to a series health hazards.
- ☆ Always clean the combine before starting as trash around the exhaust system can cause fires.
- ☆ Clean the oil, grease or mud on ladders or on the platform of the combine as they can cause serious falls.
- ☆ Check if the combine is equipped with a cab, then clean the glass to provide maximum visibility.
- ☆ Check the tire pressure of combine every day as under-inflation can lead to buckling of the sidewall, which causes dangerous tire failure.
- ☆ Check the brakes once in a week as with hydraulic brakes, it is required that the master cylinder should have full fluid and should be free from the air inside the lines.
- ☆ Check the threshing cylinder-rocking bar to see it is clear of the cylinder.
- ☆ Make sure that all shields and covers are in place and fastened securely.
- ☆ Remove or stow all service equipment.
- ☆ Always use the handrails and ladders provided on the combine for safe mounting and dismounting.
- ☆ Be sure that all PTO covers, safety stands, and shields are on the machine before taking it to the field.

Checking Bailing Hay Equipment

Crop grower should check for following before using the bailing hay equipment:

- ☆ Check that the baler should be properly connected to the tractor.
- ☆ Bailer should have required safety guards fitted.
- ☆ See that nobody should be allowed to ride on the baler.
- ☆ See that nobody is getting too close to the baler.
- ☆ Check that fire extinguisher should be fitted to the machine.
- ☆ Disengage PTO and apply flywheel brake before doing any baler adjustments.
- ☆ Stop the engine and apply flywheel brake before repairs or stringing up the baler.

10.3.2 Analysing Workplace Risk

The main causes of risk/accidents due to equipment while working at the site include:

- Transport (running over or vehicle over turnings);
- Falling from height (from fragile roofs, trees, *etc.*);
- Struck by moving or falling objects (bales, trees, *etc.*);
- Trapped by something collapsing or overturning;
- Livestock related accidents and fatalities;
- Asphyxiation/drowning.

Workplace Risk Due to Farm vehicles

Farm vehicles and implements are involved in many farm injuries and fatalities, so number of factors that will contribute to risk, will include:

- Unrestrained passengers being carried in the tray or on trailers
- Unsupervised children around moving vehicles
- Inexperienced drivers
- Poorly maintained vehicles.

Vehicle operator should always ensure that:

- Passengers use seatbelts
- The driver is trained and licensed
- Adopt safe driving methods
- Passengers are not carried in the back/tray
- The vehicle is maintained regularly
- Children are kept away from driveways and moving vehicles.

Workplace Risk due to Harvester

The hazards due to harvester include:

- Entangled with leveling or discharge of augers in the grain tank
- Falling from the machine during pre/post-season cleaning
- Contacting overhead power lines
- Being running over
- Contacting with knife, reel, stripper rotor, straw, chopper or spreader
- Being trapped under the header or injured by header falling from its transport trailer

To avoid risk don't:

- Climb or reach in grain tank unless the engine and all augers are stopped, and ignition key has been removed

- ☆ Climb around areas of the combine unless there is purpose-built access available
- ☆ Carry passengers on header unless passenger seat is provided
- ☆ Allow people to jump on or off from harvester when it is moving
- ☆ Work under an unsupported header
- ☆ Carry out maintenance or try to adjust automatic sensors with the engine running
- ☆ Park, carry out maintenance or extend the unloading auger when under or near overhead power lines
- ☆ Run the combine with guards raised or removed
- ☆ Allow children on or near the combine
- ☆ Brake harshly, especially on slopes.

Harvester operator should:

- ☆ Make sure that anyone working on harvester has read and understood the instruction book
- ☆ Carry out a 'Safe stop' procedure before carrying out maintenance
- ☆ Gave a system for thorough preseason maintenance checks to ensure that all brakes, guards and other safety devices are ineffective working order
- ☆ Use any grain sampling mechanism provided
- ☆ Have a safe system of work for combining paddocks where there are overhead power lines.
- ☆ Use appropriate farm maps as reminders when daily work is being planned
- ☆ Agree on routes for travel with a local company which take into account safe clearance distances
- ☆ Follow correct procedures for working under the header or when transferring header on and off its transport trailer
- ☆ Provide and use correct tools and knife guard to aid safe removal, handling and storage of the cutting knives
- ☆ Work with cab door shut
- ☆ Be particularly careful when reversing
- ☆ Wear slip-resistant footwear and avoid wearing loose clothing
- ☆ Take care on steep ground and avoid sudden changes of direction.
- ☆ Park across slopes if possible and ensure the handbrake is fully applied before leaving the vehicle

- ☆ Make sure all guards are in position and correctly fitted before starting work
- ☆ Regularly clean straw and chaff deposits from the engine compartment and around belts and pulleys
- ☆ Make sure there is suitable fire-fighting equipment available on the combine

Glossary

Accreditation: The formal recognition by an independent body, generally known as an accreditation body, which is a certification body operates according to international standards.

Applicant: Candidate who applies for GAP certification by an approved certifying body.

Assessment: An appraisal of procedures or operations based largely on experience and professional judgment.

Audit: The Organization for Standardization (ISO) defines an audit as a systematic, independent and documented process for obtaining audit evidence and evaluating it objectively to determine the extent audit criteria are met.

Audit Client: Any person or organization that requests an audit. The Client can be internal or external.

Auditee: Entity or organization (or part of an organization) that was, or is, being audited.

Audit Evidence: All the information collected during the course of an audit, which serves as the basis for the auditor to make an opinion and determine compliance with the requirements (standard) being audited against. Such evidence includes records, factual statements and other verifiable information (*e.g.* observation of work activities and physical examination of products, materials and equipment) that is related to the audit criteria being used. There must be sufficient audit evidence for the auditor to submit a final opinion. Audit evidence can be either qualitative or quantitative. Objective evidence is information that shows or proves that something exists or is true.

Audit Findings: Result from a process that evaluates audit evidence and compares it against audit criteria. Audit findings can show that audit criteria are being met (conformity) or that they are not being met (non-conformity). They can also identify best practices or improvement opportunities.

Audit Itinerary/Schedule: An audit itinerary is a schedule or timetable of activities to guide the on-site audit process, which will allow for efficiency and time management, although deviations from the timetable may occur sometimes.

Auditor: An official trained and qualified to conduct an audit on behalf of a certifying body. In the GLOBALG.A.P. system, an auditor conducts an audit of the quality management system where applicable and can also conduct inspections of production sites.

Audit Plan: An audit plan specifies how a particular audit will be conducted. It describes the activities that will be carried out in order to achieve the audit objectives.

Audit Report: A standardized means of reporting the audit findings and non-conformances (exceptions) with respect to the appropriate level of management.

Audit scope: A statement that specifies the focus, extent, and boundary of a particular audit. The scope can be specified by defining the physical location of the audit, the organizational units that will be examined, the processes and activities that will be included, and the time period that will be covered.

Biodiversity: The variability among living organisms from all sources, including 'inter alia' terrestrial, marine and other aquatic systems, and the ecological complexes of which they are part; this includes diversity within species, between species and of ecosystems.

Calibration: Determination of the accuracy of an instrument, usually by measurement of its variations from a standard, to ascertain the necessary correction factor.

Certification: The provision by an independent body of written assurance (a certificate) that the product, service or system in question meets specific requirements.

Certification Body: A third party auditing organization that audits facilities against a specific international standard or code. The GLOBALG.A.P. definition is an organization that provides conformity assessment services such as inspections and certifications to producers and producer groups against the GLOBALG.A.P. standards in accordance with ISO/IEC 17065 accreditation requirements, GLOBALG.A.P. General Regulations and License and Certification Agreement.

Checklist: An inspection and audit tool with documented questions that reflect the requirements, procedures, or policies of an organization. For GAP inspections/audits it can be used by producers, producer groups, certification bodies or organizations which help producers to implement GAP standards towards obtaining certification (or GLOBALG.A.P. certification).

Compliance Criteria (CC): Information provided to further illustrate each control point and how to successfully address the requirement(s) identified in the control point.

Conformance: Meeting or complying with established requirements as set out in established procedures or of a particular standard.

Control Points (CP): Each of the requirements requested by a standard (or GLOBALG.A.P. standards) to implement good agricultural practices. Within the GLOBALG.A.P. standards, control points are classified as Major Musts, Minor Musts, or Recommendations.

Control Points and Compliance Criteria (CPCC): The comprehensive set of control points and compliance criteria that define the standard against which a producer's performance is measured both internally and externally.

Exception: That which is a deviation from a standard. Exceptions are those items that do not conform to general audit finding(s).

Finding: An auditor's (or audit team's) opinion as to the farm's/facility's overall GAP performance during the audit period. The finding is based upon evidence gathered during the audit process.

Food Safety: The assurance that food will not cause harm to the consumer when it is prepared and consumed according to its intended use.

Good Agricultural Practices (GAPs): Practices that address environmental, economic, and social sustainability for on-farm processes, resulting in safe and quality food and non-food agricultural products.

Hazard (GAPs): A biological, chemical, physical or any other property that may result in a situation that is unsafe for workers, consumers, or the environment.

Hazard (food safety): A biological, chemical, or physical agent that could contaminate food at any stage and cause an unacceptable health risk.

Hazard Analysis Critical Control Point (HACCP): A food safety system that identifies hazards, develops control points throughout the flow of food, sets critical limits, and monitors the effectiveness of these control measures.

Internal Controls: The various engineered and managerial means-both formal and informalestablished within an organization to help the organization direct and regulate its activities in order to achieve desired results; they also refers to the general methodology by which specific management processes are carried out within an organization.

Mass Balance: A reconciliation of the amount of incoming raw material against the amount used in the resulting finished products, taking into account process waste.

Non-conformance: Activities carried out that are not according to the established procedures or requirements of a particular standard. A non-conformance may be minor or major.

Observation: A statement about something that has been noticed.

Quality Management System (QMS): The organizational structure, procedures, processes and resources needed to implement quality management.

Record: A document containing objective evidence illustrating activities being performed and/or results achieved.

Risk: The chance that a condition or set of conditions will lead to a hazard.

Risk Assessment: An estimate of the probability, frequency and severity of the occurrence of a hazard.

Sample/Sampling: Selecting a portion of a group of data in order to determine the accuracy or propriety or other characteristics of the whole body of data.

Self-Assessment: Internal inspection of the production system and the registered product carried out by the producer or a sub-contractor, based on GLOBALG.A.P. checklist (or checklist from another GAP scheme).

Standard: A document that provides requirements, specifications, guidelines or characteristics that can be used consistently to ensure that materials, products, processes and services are fit for their purpose (ISO).

Substantiate: To establish or prove conclusively.

Traceability: The ability to retrace the history, use or location of a product (*e.g.* origin of materials, processes applied or distribution or placement after delivery) by means of recorded identification markers.

Verification: Confirmation by examination of evidence that a product, process or service fulfils specified requirements.

Worker: Any person or a farmer who has been contracted to carry out a task. This includes farm owners and managers, as well as family members carrying out tasks on the farm.

Index

Q

R

S

9 789354 615092

Printed at Repro India Ltd.